The Diamond Drilling Industry

Hancock House Publishers

DAN FIVEHOUSE

The Diamond Drilling Industry

ISBN 0-919654-69-X

Copyright © 1976 Dan Fivehouse

Canadian Cataloguing in Publication Data

Fivehouse, Dan, 1946-
   The diamond drilling industry

    Includes bibliography.
    ISBN 0-919654-69-X

      1. Rock-drills. 2. Boring. 3. Mining industry and finance - Canada - History.
I. Title.
TN281.F59   622'.24   C77-002020-8

All rights reserved. No part of this publication may be reproduced, stored in a retrieval system, or transmitted, in any form or by any means, electronic, mechanical, photocopying, recording, or otherwise, without the prior written permission of Hancock House Publishers.

Designed by Nicholas Newbeck Design

Printed in the United States of America

HANCOCK HOUSE PUBLISHERS

3215 Island View Rd.
Saanichton, British Columbia V0S 1M0

12008 1st Ave. S.
Seattle, Washington 98168

Contents

FOREWORD	7
INTRODUCTION	15
THE FILE ON FELIXA	21
A SHORT HISTORY OF DIAMOND DRILLING	75
ILLUSTRATIONS	81
DIAMOND DRILLING IN WESTERN CANADA	125
SOURCES	191

Foreword

At the turn of this century George Alfred Denny authored a book entitled DIAMOND DRILLING FOR GOLD AND OTHER MINERALS. The diamond drill as an instrument of Industrial Revolution technology was not quite 40 years old. Denny's book was perhaps the first text in the English language dealing with this new field of expertise. The book was essentially a handbook for drillers, part of a series of technical books issued by his English publishers. Denny, a British mining engineer working in the gold fields of the Transvaal of South Africa, reduced the subject to its basics in his preface. Much of what he wrote in 1900 remains familiar today:

The business of gold mining is being rapidly evolved from the sometime blindly speculative to a legitimate commercial enterprise, in which it is clearly recognized that the increased profit potentialities compensate for certain added risks which are involved in its prosecution. From its very nature it is frequently impossible with any degree of certainty to estimate what results may be expected from a given mining venture... The uncertainties are primarily (1) the average milling grade of the ore; and (2) it width and extent. The manner of settlement of these has too often in the past been upon lines which are not to be recommended in any way...

The diamond drill bores a perfectly smooth hole to any depth or in any given direction from vertical to horizontal, bringing to the surface a solid section or "core" of all strata passed through, in the order in which they lie, determining also the exact depth of any particular rock, its thickness and characteristics. The size of the core...allows of thorough examination and testing, and the presence or otherwise of the ore or mineral sought is settled beyond a doubt. It also gives positive information of the rock that will be met with in the shaft sinking, thus making it possible to closely estimate the cost of such work.

Diamond drilling remains very much an *information service*, providing the basic data for geologists and engineers and cost accountants to work with. The need for precise information about the earth structures underlying a project area remains of critical importance whether the project involves minerals exploration or mine development or the construction of dams and bridges and high-rise buildings. Diamond drilling continues to provide this information in a most precise and economical manner.

As we shall see, the diamond drilling industry concerns itself with a relatively limited but very technical set of reference points: linear feet drilled, dollar cost per foot of drilling, percentage of usable core retrieved, percentage deviation from true of the drill hole and so on. It is detail work, but detail work carried out in locations and under physical conditions which constitute some of the most hostile and strenuous working environments to be found in any occupation. The peculiar requirements of this work—physical and mental toughness, technical expertise and special problem-solving abilities—set diamond drillers apart as a very curious breed of craftsmen. The pressures of work often inspire drillers to develop as very individual—and colorful—characters.

In his 1967 book, THE DRILLING OF ROCK, K. McGregor opens his chapter on diamond drilling with this observation:

The diamond driller must be considered the aristocrat of the drilling world. Whether working a small portable rig or engaged in oil-well prospecting, he must be resourceful, ingenious, patient and systematic. His equipment is costly, and the hole he produces is worth a considerable amount of money.

Unless you are a neighbor of a diamond driller or come in contact with drillers in your profession, chances are that you know little about this fascinating industry aside from the assumptions that common-sense would have you make. Aside from technical papers and a few industry-sponsored drilling handbooks, there is virtually no information available to the general reader concerning the diamond drilling industry.

Indeed, you would be hard pressed to find mention of the subject in your favorite encyclopedia. It came as no surprise in researching this book that librarians and archivists responded to my information requests with replies ranging from: "Aren't those the drills they use to discover diamond deposits?" to "I thought all drills had diamond tips" and "I didn't know we had a diamond industry *here*!". The confusion is natural enough and this book seeks to bridge that information gap in however small a way.

A survey of diamond drilling literature reveals that the next mention of the subject after the publication of Denny's book in 1900, comes in the book ROCK DRILLING written in 1912 by Richard T. Dana and W.L. Saunders. Published in New York by John Wiley & Sons with the assistance of the Ingersoll-Rand Company, it was an investigation into the economics of blast hole drilling and served as well as a promotion for Ingersoll-Rand drill products. It includes mention of diamond drilling in its coverage of various aspects of rock drilling.

K. McGregor's THE DRILLING OF ROCK, which was sponsored by the Chicago Pneumatic Tool Company and the Consolidated Pneumatic Tool Company, opens with a quotation from the Dana/Saunders book and again includes mention of diamond drilling in its more general coverage of rock drilling.

It was not until 1956 that a second book devoted exclusively to diamond drilling made its appearance. J.K. Smit & Sons Diamond Products Ltd., a major industry supplier, sponsored James D. Cumming's DIAMOND DRILL HANDBOOK which is updated regularly with the assistance of A. Percy Wicklund. This book has become something of a 'bible' for the trade and rightly so, for it covers topics as diverse as the production of raw diamonds to samples of forms and paperwork to be used for contracting services.

All four books, Denny's included, share this in common:

they are all essentially handbooks for drilling personnel and supervisors. This closed-circle of diamond drilling literature provides any writer approaching the subject with a boon and with a handicap. On the one hand, here is an entire industrial field open for exploration and a writer need not be too concerned about treading on more venerable toes. On the other hand, the field of information lying ready to be mined is so vast that it is difficult to decide where to begin. There is the further consideration of leaving a receptive and knowledgeable audience for future writers to discover, and yet providing those future scribes with sufficient room to move about and explore the subject from other viewpoints. For these reasons, this book does not attempt to lay claim to any degree of completeness with regard to the treatment of the subject matter of diamond drilling.

The focus of this book is on Western Canada and on the major diamond drill contracting companies that had their beginnings in this region. Mining in the West, especially in the Kootenay and Boundary districts of British Columbia, came of age during roughly the same period of time that the diamond drill began to emerge as the 'aristocrat' of power drilling technology. The Kootenays became something of a center for diamond drilling technology, providing the fledgling industry here with a proving ground for its techniques and hardware, and offering the diamond drillers a challenge in the rock-solid form of the infamous Kimberley *chert*—said to be the hardest natural rock structure in the world. Three of the great contracting companies to emerge from the Kootenays and surrounding areas were the Diamond Drill Contracting Company, Boyles Bros. and the T. Connors Diamond Drilling Company Limited.

The contemporary Connors Drilling Ltd., a subsidiary of Bow Valley Industries based in Vancouver, can trace its ancestry through all three of these early firms. Celebrating its 50th anniversary in 1976, the Connors firm became curious to know more of the details of that time around the beginning of this century when Rossland and Nelson and Kimberley and Kaslo were glittering names in mining circles. A series of short magazine articles did not do justice to the subject and the company generously decided to commission this book. In addition, the company arranged introductions for the author with the old-timers in the area who were themselves part of this moment of history, with company employees and industry sources who provided me with invaluable material, and opened company records to assist me in my work.

There will inevitably be pieces missing from my puzzle and these fragments must now wait for another time. I have tried to present a well-balanced text for the general reader, knowing that those familiar with portions of my story will bring even richer memories to the reading of this book. In order to acknowledge all the assistance that I have received during the course of my work I have included a SOURCES section following the main text. Interested readers may wish to use SOURCES as a door to further explorations of the subject of diamond drill contracting.

And just what is diamond drilling all about? Let's look...

Dan Fivehouse
Vancouver, B.C.
1976

Introduction

Diamond drill contractors today provide a wide variety of services for many different industries, institutions and government agencies: soil sampling, cone penetration testing, blast-hole drilling, mine development and foundation testing for the construction of dams and bridges and wharves and high-rise buildings. Diamond drilling first came to prominence as a technological innovation in the field of minerals exploration and exploration drilling remains the most challenging and, yes, the most romantic pursuit of the diamond driller. Let us begin with a look at the modern-day activities of diamond drill contractors, with our focus on exploration drilling.

Money! For better or worse, money is the operative word in our Twentieth Century vocabulary. Money has always been a first object of concern for the mining industry. The first organized explorations for mineral deposits probably occurred when mankind sought out those few varieties of metals which could be used to fabricate tools and weapons and jewelry. Pieces of metals such as silver, gold, copper and bronze came into use as barter items and soon assumed a significance apart from their possible conversion into fabricated items. They acquired a symbolic status as it became widely understood that they could be traded for items of more immediate use: foodstuffs and clothing and so on. They became coinage—the cornerstone of the concept of money—and over a period of time they became standardized in size and weight and value. The value of particular coins was regulated within social groups by the imprinting of symbols upon their surfaces.

Eventually the value of coinage had more to do with those symbols than with the possible use of the metal itself. Our modern concept of money has evolved to the point where many of the markers that we use to replace the awkward coinage (paper currency, stocks and bonds, savings certificates, etc.) deal no longer with things of value, but with promises of worth. Some of these promises of worth extend far into our personal futures, so money has come to assume an element of time as a boundary of its definition. We are beginning now, in our society, to value things more and more in terms of their relation to time. Indeed, we have now begun to deal with time as a commodity—if only in a figurative sense. We sell computer *time*, consultants sell their personal *time* and so on. Money has become equated with *time* and there is a rightness to that marriage.

In certain industries time has always been observed to be an adjunct of money: in the realm of agriculture, crops have to be harvested and transported to market and sold within a certain period of time or they lose their value in a most dramatic way. The mining industry has always played an intimate role in the world of 'money' by providing us with the actual minerals and metals for the minting of our coinage. What is less generally recognized is that the mining industries must always include *time* as a vital component of their internal economics. Mineral deposits are regarded as 'wasting assets' which will have an ultimate value of zero once they are used. All of the costs of developing mineral deposits, including projections of market performance over an established period of time, must be calculated in advance to ensure that the mining company will be able to operate the mineral deposit in good economic health.

The mining industries are involved in the selective extraction from the earth of particular metals and minerals, oils and gases, which form the backbone of our 20th Century lifestyles, in response to society's demand for these resources. The costs of locating and extracting and refining these commodities is very high. The costs have now become so great that the large sums of money used for exploration and development are generally beyond the grasp of individuals acting alone. These 'capital' sums of money can only be generated through extensive communal organizations such as governments, banks and public corporations.

There are always certain risks involved in committing large

sums of capital to develop mineral resources, and governments rarely assume those risks except in times of national emergency when the strategic need for certain products outweighs the costs involved. Instead, most governments leave the development of mines and oil fields and related resources to the private sector and content themselves with exercising some degree of control over these industries by securing loans and granting tax shelters, balancing profits and prices through taxation and royalty payments, and monitoring worker safety.

The private companies that develop mines are composed of and answer to groups of individuals who are willing to commit money to these ventures for particular reasons. Most often, they are looking for a way to make their personal wealth grow in volume, to make a profit. The costs of developing a mining property must be structured in advance to provide that profit relative to the competition in world markets.

Mining companies require a great deal of very precise and sophisticated information to allow their executive decision-makers to decide: 'Do we proceed with work on this property? If so, when and how do we do it? Where do we get the money to finance this project, how do we use the money and what can we promise investors in return for its use?'. The information required includes data as general as knowledge of the basic geological structure of the region for miles around the site and as specific as samples of ore taken from the heart of the mineral deposit. This information must be purchased at great cost in terms of time, energy and cash. It is certainly not purchased as a gift for a company's competitors, so it is treated with the utmost secrecy by the people who research, analyze and use it as the basis for their decision-making. For this reason, the exploration sector of the mining industries has become one of the most secretive business groups in the modern world.

The cost of the expertise required to research and evaluate resource development has become so expensive and so specialized that few companies can afford to maintain these experts on their payrolls full-time. Most mining companies now contract these portions of their exploration work to independent firms who must maintain a high level of secrecy and provide results with a very high percentage of accuracy. These consulting companies usually contract the actual physical work to yet other independent firms, like diamond drill contractors.

The most basic and perhaps the most important piece of raw data examined by the mining companies and their consultants is the core sample taken from the mineral deposit. The man responsible for extracting that sample from the earth, intact and in usable form, is the diamond driller.

Once a mine is developed and producing, the drilling budget for all aspects of exploration and development remains relatively stable at about 2-5% of the total cost of operations. In the exploration phase, the diamond drilling can account for more than 50% of the budget and there is enormous pressure exerted on the diamond driller to produce the most accurate results humanly possible. In today's market, diamond drill contractors compete on a basis where their success or failure to win contracts depends on their ability to recover perhaps 93% usable core at a stated price per foot, rather than a competitor's 92% recovery for the same price.

In order to understand something of the conditions under which the work is done, to understand the importance of the diamond driller's work with regard to the decisions made within the exploration phase and to appreciate the subtleties of the driller's life, let's join a project undertaken by an independent contractor to perform exploration drilling in a remote area of the globe.

The File on Felixa

I

The international headquarters of CANADA COREX LTD., diamond drilling contractors, occupy the fifth floor of a solid, older building overlooking the Vancouver waterfront. A stone building with certain Victorian airs, such as the stained-glass panels of the stairwell windows, the office complex had recently been gutted to the 12-inch beams and renovated by a firm of young architects hustling to make a name for themselves in urban redevelopment. The man in the spacious office at the back of the fifth floor was sympathetic to their aims. That was one reason why he had moved COREX to these premises when the time had come to expand.

One other reason that he decided to move here was for the view. The old offices farther downtown, near False Creek, had faced south. He much preferred the north view containing the entire spectacle of the busy inner harbor from Stanley Park on the west to the shipping terminals surrounding Centennial Pier on the east. This view of western Canada's major port was limited on the north by the massif of the North Shore Mountains which remained snow-capped for much of the year.

From this office, near the foot of Granville Street, the man could sit and survey the scene and think and plan in more congenial surroundings than before. Below him, near Brockton Point in Stanley Park, various pleasure boats gassed up at the service floats and the float planes took off and set back down in their daily dance around Coal Harbour. Just now he saw the yellow and black Beavers of Tyee Airways taxi out into the harbor like giant amphibious bees. A red and white Twin Otter of Air West made its landing approach. He knew them all, by color and species. COREX was a favored customer of the small air services. For years before the first foreign contracts brought international recognition for the firm, COREX drillers had punched holes in most of the known mineral deposits on the British Columbia Coast and in

23

the north where the main sources of transport were the float planes and helicopters. COREX remained one of the most active contractors serving the remote areas of Western Canada.

The man in the spacious office was alternately musing about those float planes and their connections with COREX jobs under way up the coast, and thumbing through a large stack of magazines and newspapers which perched just on the corner of the desk. He was not himself behind the desk. Rather, he had drifted halfway across the room on his swivel chair following the view along the expanse of windows, moving with the sun. He was sprawled in the chair with his feet up on the windowsill thinking how seldom he ever got to sit still and admire the view after all. He had returned just that morning from an inspection trip up north and he was to leave again sometime soon for an industry conference in Montreal.

After two hours of dictating replies to various correspondents, he had decided to take a few minutes to catch up on his professional reading. He skimmed much of the material in the pile, reading a bit here, a bit there, then went to work with a very small, very sharp knife expertly slicing selected articles and news items from the glossy pages. The pieces he selected all had an immediate bearing on projects of interest to him or projects which showed future promise. The exhausted magazines he dropped to the floor beside him, building an impressive monument of shredded paper in the middle of the office.

After working his way through the *Western Miner*, *The Northern Miner*, *Oilweek*, the *Journal of Commerce*, *Financial Post* and two weeks accumulation of various other U.S. and Canadian periodicals, he carried his clippings to his desk and began making notes on a small collection of file cards which he passed on to his executive assistant as leads for possible contracts to pursue. He was busy sorting through this material and thinking about the next year's prospects for work when the telephone rang through into his office.

"There is a call from Harry Demarche, Mr. Dixon," Ruth informed him.

David Armstrong Dixon, COREX general manager and resident handyman, asked his secretary to connect the caller.

"David?"

"Harry! Well, it's good to hear from you, where've you been?"

"Traveling, same as yourself, I hear."

"Yes, well, just drifting around the country checking up on the jobs."

"Ah, it's good to hear they're getting some work out of you!"

"Is it?! Well, what can I do for you this fine day—or don't you have enough change to buy lunch?"

"Lunch sounds good, I accept," Harry chuckled, "but not today. I'm booked out all afternoon, but how about tomorrow? I called to ask if you were interested in bidding another job in Felixa."

"Felixa again?"

"Yes, and there's a bit of a rush, so we should talk about it as soon as possible. If you are free tomorrow morning, I'll have Sam Ellis here to explain the technical aspects of the work to your people and then maybe we could have that lunch?"

"Sounds good to me. Your place? What time?"

"Oh, about ten o'clock? It's pretty straight forward, so I don't suppose you need to bring the shop in on it yet."

"I'll bring George Bryant with me, and Georgina. Norm and Jack Chang are still out of the country."

"Good, see you then!" And with that, Harry disconnected.

    *   *

In the bright mid-morning of April 10th, on a beautiful spring morning following five months of rain, David and George and Georgina Laing walked up Granville Street to the Pacific Centre and entered Vancouver's new shopping mall/office complex off Dunsmuir Street. David stopped at the United Cigar Store to stock up on the petit Dunhill *Chicas* that he favored, then met the others window shopping on the upper level mall. Together they strode through the bright shopping area which was relatively deserted on this fine spring day to enter the lobby of the black glass tower that housed the offices of Manning, Demarche and Barrett Ltd., consulting engineers.

They were ushered into the boardroom that Harry had reserved for their briefing and renewed acquaintances with the MDB people. Besides Harry and Sam Ellis and his assistant, a 'junior rock doctor', there were two others present. The MDB company travel coordinator and chief expediter, Frank Burdon, was there to explain the company's arrangements with the Felixan government for work permits and the like. Chapel Hendry, MDB's chief of accounting services was present to explain the financial requirements of the bid to Georgina Laing.

Harry Demarche gave them all a brief outline of the job and of MDB's role as consultants to the Middle Eastern government:

"It's an interesting job, I'll give you that much from the start, and we thought COREX might like to take a crack at it since you have such a splendid record in the handicap race with Arab cooking. We'll be talking with a number of other contractors—all of your competitors, in fact—but in a way you already have something of an edge that we've given you so you get first shot at following it up. I don't remember how much you know of the situation in Felixa, so forgive me if this becomes repetitious. There's a bit of what I have to say that will be new to you, I think. At one time the entire region, of which Felixa is now a small subdivision, was one of the wealthiest Middle Eastern empires—but not anymore. For the past 1,500 years, the people have lived in relative poverty, because the country's resources were not of immediate concern to the rest of the world. The old empire, based on agriculture, fell to pieces when the world's trade routes were reorganized at the very beginning of the modern era. The peculiar geographics of the area placed the Felixan highlands just far enough out of the way of the major trade routes to become bothersome. But today, the Saudi Arabian peninsula is alive with activity and everyone is cashing in on natural resources like oil and gas and what have you—except Felixa. For too many years, the local hierarchy discouraged foreign activity in the country. Now, after the republican take-over, Felixa sincerely wants to invite development capital but foreign investors have taken a 'wait-and-see' attitude.

"There is no oil in the country to speak of and throughout North Africa and the Middle East there are proven reserves of a great many metals and minerals, so why should anyone bother with Felixa where history has only shown occasional traces of mineral wealth? This is a question that we have been hired to answer with enough positive fervor to attract development money to Felixa. We feel that there is enormous promise for a healthy mining industry in this country, and now we must prove out our theories. If there is indeed enough mineral wealth here to support our expectations, then we will get first shot at developing a mining industry for Felixa.

"But, there is a timing factor involved. There are a number

of other national groups currently working in the country to develop industry of one sort or another and we have a suspicion that if we don't produce results in a reasonable period of time, we will lose the initiative to the Germans or the Russians or the Americans or the Chinese—all of whom are keeping close tabs on everything that happens there.

"Our foothold in Felixa is our proposed development of the salt dome at Salif and you, of course, know more about that than anyone else since you are our exploration contractor at that site. The new government has been watching the work at Salif and they are very impressed by what we have accomplished so far—by what *you* have accomplished. I understand that Pat Delise and the boys gave the governor a guided tour when he was in the neighborhood and he was very impressed with diamond drilling. A very smart man, the governor—enlightened and indicative perhaps of the new generation in charge there. He made a very forceful presentation on our behalf to the Presidential Council and the government came back at us with feelers about launching a mining industry for them.

"Essentially, we are to follow-up reports of mineral discoveries that are well-documented in historical records. If we find anything economically significant, we have an option to try syndicating its development. The government of Felixa needs to establish some record of credibility with the rest of the world in order to attract the huge sums of money they will need to cope with the world in decades to come. And they need to make allies to help them keep their balance in the shifting political currents of the Middle East. They are serious enough to risk some of Felixa's dwindling cash reserves as seed money for this exploration project and if we prove up reserves, they will smooth the way for a consortium of foreign companies to participate in their economy.

"You helped us sell them this concept with your work at Salif. That's why we're giving you first notification of our call for tenders, and one of the reasons we are meeting today is to give you some background on the economics of this thing. We have to bring in something to work with fast and on a tight budget. If and when we begin mine development, there would probably be more drilling work with more money available at that time. But we're very tight for now.

"As for time, there's perhaps another week or so of rain before the dry season begins. Then three months of good weather before the big rains drench the land from July until September. We'd like to have something to show the government before the monsoons—figure maybe a month to organize this and begin the work?"

There were gasps in the room. Harry grinned from beneath heavy eyebrows.

"I wasn't kidding when I said things were tight—time *and* money..."

"It takes more than a month to mobilize anything overseas," protested George Bryant.

"Before—maybe. Not this time. Your bigger competitors have certain advantages over you: manpower, economics, and so on. This is something of a long shot for you—I realize that—but, well, you do have a crew that knows the country and you've got a day or two advance notice to prepare your bid. There's not much more that we could do—we only put this deal together last weekend. Now I'll let everyone have a minute or two to outline our project and then we'll get down to specifics. Sam?"

After the briefing, Chapel Hendry and Frank Burdon took Georgina Laing to their section to review costs. George Bryant went with Sam Ellis and his assistant to the geology section to discuss rock structures, and David Dixon took Harry Demarche to lunch.

* *

David was back in his office by mid-afternoon. As he came into the COREX reception area, he was hailed by his personal secretary, Ruth Thompson.

"Mister Dixon, some calls for you!"

David nodded for her to proceed.

"Mister Harrington and Mister Chang called to report that they have reached Kuala Lumpur and they're meeting today with representatives of the East Asian Syndicate. Norm said that he will call tomorrow in the morning to brief you on the talks."

Norman Harrington and Jack Chang were the COREX roving project consultants charged with developing future projects. Jack was the COREX financial wizard and chief purchasing agent. The two men, traveling as a team, followed up all the personal contacts and business leads developed by COREX personnel. It sometimes appeared that Norm and Jack logged more airline flight hours in these pursuits than the rest of the company combined. At the moment, they were working to arrange an exploration contract with the East Asian Mining Syndicate for a two-year drilling program in remote regions of Burma, Thailand and in Malaysia. David was pleased to hear that progress was being made on that front.

"Margaret called and asked if you could call her back whenever you get a free moment. George Bryant said he'd be back in the office for just a short time this afternoon and he wanted to see you then for a minute or so."

"Good, I want to talk to him, too. Just send him in when he arrives," David replied and then walked off into his own office to return his wife's call.

George Bryant knocked on the opened door while David

was on the phone speaking to Margaret. David threw him a little salute and George disappeared. He was back within minutes with two big mugs of coffee, just as David replaced the telephone receiver on its cradle.

"Ah, thank you," David said as he took the mug by its handle and swept it up from the desk to drink. George relaxed in the Scandinavian-styled lounge placed near the windows on the other side of the desk. There was music playing very softly in the background. David Dixon was a compulsive listener and he always worked with a backdrop of music in the room. He appreciated the current of logic that flowed through music and it helped him to think. For a few minutes, the men sat and drank their coffee and listened to the music.

"So, what do you think?"

"Well, I think we can put a bid together on time," said George, "but I do wish that Norm and Jack were back and that we had Pat Delise here for the next few days."

David nodded assent: "Yes, that would be ideal. But we're just going to have to lay responsibility on Georgina and let her handle it alone—the paperwork anyway. It would be nice to have someone here to oversee the preparations who will be heading for Felixa if we win our bid. I'm going to move Pat up from Salif to organize the job in the highlands, but I'm not sure who else to send. Got any suggestions?"

David waited for a reply. He knew that George would answer very carefully. George was always one to offer an opinion that covered as much ground as possible and which he considered to be in the best interests of the firm. It was right that these two men should make these decisions, because they were two of the five principal shareholders in the firm. Both Norm and Pat were overseas and Edgar McGavin, who ran the COREX shop, preferred to leave staffing problems to the front office. Many of COREX's 150 employees held shares in the company, but these five constituted a majority interest and were listed as company directors. Whatever decisions they made were reflected in the cash dividends paid to all the other shareholders and they were all very careful with their logic.

"Who else is at Salif?" asked George.

"Billy MacDonald, Gordy Watson and Mike Perrone."

"And who would you leave to finish Salif?"

"Well, Gordy Watson is senior and a pretty good man. I'd go with him."

"You might have a problem there, Dave," George spoke softly, settling back in his seat, "I don't know if you've heard, but Gordy's wife just kissed him off."

"No! When did that happen?"

"Sometime in the last couple of weeks. Angie heard through the grapevine that Marie sent him a 'dear John' washing her hands of things—and then took off, disappeared. Nobody has heard from her since. I gather you haven't had news of this?"

"No, I hadn't heard. I wonder if Pat knows?"

George shrugged, "With mail service over there, I wonder if Gordy knows. I don't suppose it matters much—Marie was his third wife, wasn't she? He's been through it before."

"Damn it! I wish I'd known—I would have had him sent back until they got things straightened out..."

"What use?" George consoled, "He volunteered for Salif— just about jumped at the chance as I remember, so he must have wanted to get away pretty bad. Maybe he knew it was coming."

"Yeah. I guess so..."

David was silent for a minute, tapping his fingertips together and looking out from behind his desk with his face set in a frown. He had seen this happen so many times now, with Gordy and with others—it was the major occupational hazard in the emotional life of diamond drillers. And now it threatened to become a problem for him. There was no telling what a man might do under stress ten thousand miles from home, and the last thing they needed now at Salif was any interruption of the highly successful drilling routine. Besides that, Gordy Watson had once had a drinking problem...but no, that would be no problem there, David reminded himself, best place he could be was in an Arab country where no booze was allowed. He sighed.

"Well, for now I'd say I'll stick with Gordy to take over Salif."

George nodded: "So that's Salif..."

"Yes..."

"All right, what about the highlands? I guess you might want one or two of the younger drillers there to grow up with the job?"

"Yes, that would be good."

"I've got a live one for you: Ken Holby. He's working the Yukon right now at Pelly River. He's maybe thirty, has a way with people. Gets along fine with the Indians—and the tourists—and we've had a really smooth time of it ever since we moved him there. He's got five or six years under his belt as a driller, a few more as a helper, and he's single."

"Good! Can you spare him?" David asked of his project manager for North America.

"Yes, he's a fine teacher as well and he's got two helpers there who are just about ready to move up as drillers. I was supposed to fly up north tomorrow morning to estimate this summer's work at Great Bear Lake. I don't know if I should put that off?"

"No, if you can do it in time for the weekend go ahead. That is an important bread-and-butter contract and we don't want to look careless. Go ahead, but be back at the beginning of the week, so we can begin preparing this other bid."

"Right. On my way up to Norman Wells, I'll make a quick stop at Pelly and drag Ken away—providing he thinks the others are ready. I'll send him down to help with the bid and the mobilization. He should be here by the time I get back."

"Thank, George. And if you can think of some more names for me, call from Whitehorse," David reminded him.

* *

The telephone rang. It was not quite noon on Thursday, April 11th. George Bryant was on his way to Whitehorse and points north and David was busy assembling the information they would need to bid the exploration job in the Felixian highlands. He leaned across his papers to retrieve the phone from its precarious position at the far edge of the desk.

"Mister Dixon? Mister Harrington is calling from overseas, sir."

"Thank you, put him on," David grabbed a pen and note pad.

"Hello? Dave?" Norm's voice wavered on the line.

"Here! How are things going?" David shouted at the phone to make himself heard half a world away. After a brief pause,

34

Norm began his report. The crackle and static that normally accompany overseas transmissions threatened at times to terminate their call.

"Goddamn telephones anyway!" David anguished as Norm's voice floated away for the fourth or fifth time, "Listen Norm—are you there? OK, we just got an invitation from Harry Demarche to bid another job in Felixa. Yes. In the highlands up near Sana. Looking for copper or anything else interesting," and David explained the deal to his assistant. When he was through, he asked: "Have you got any suggestions? What? Yes, I'll hold..."

Norm was back on the line within minutes: "David? Jack has been talking with some of the Syndicate's technical reps. They were pumping him for information about the Wesdrill. Seems they saw one last week in Germany—at the heavy equipment show at the Hanover Fair. Maybe it's still there. That's more than half-way to Felixa. With that rig, we would only have to field a crew about the same size as Pat's bunch at Salif and we could use a lot of the data we researched to bid that job. Plus we'd come up with an impressive production record in a relatively short time. Jack says he was thinking of swinging a deal on one of them this summer anyway, maybe we could make an arrangement on the machine in Germany. Sana's got a good airport, eh? We could fly it in, line up a CAT and move it right into the mountains..."

"Fantastic idea! Tell Jack he's got it! That's a hell of a good plan. Listen, I'll let you go now, but check back here before you come home. Yeah, take care and good luck!"

David replaced the receiver, sketched some figures on his note pad and was back on the telephone talking to Wesdrill Manufacturing in the Vancouver suburb of Richmond.

*   *

Exactly one week later, on Thursday April 18th, five very weary people staggered into David's office to collapse in the chairs and on the couch and put up their feet to relax. The accounting staff and the headquarters office staff were gone home to dinner, and those gathered there hoped their turn might come to leave very soon. For most of the past week they had worked non-stop to prepare the COREX bid for Manning, Demarche & Barrett. Some of them had worked sixteen hour days, trying to anticipate every contingency, every expense that the company might incur as an agent of the consulting firm. Two secretaries had spent much of the previous day typing the bid and it had been delivered just before five o'clock Wednesday night to Harry Demarche. All day today these people had worked to catch up with their regular routines.

David was absent from the office the whole day, but he had called in at mid-afternoon to ask all of them to stay late to wait for his return. There was no reason they couldn't relax while they waited. Ruth was sitting on the couch talking with Georgina Laing, Jack Chang's assistant and the company's other financial wizard. Ed McGavin, supervisor of technical services and head of the Vancouver shop, was speaking with George Bryant and Ken Holby about the quarter-final match-ups for the Stanley Cup playoffs that were just about to begin. All of them were drinking cold beer that Georgina had found in the fridge of the office kitchen. Everyone was deathly sick of coffee by now, and starving. A local FM station spilled soft rock music into the room.

David breezed in at five-thirty with three shopping bags full of delicatessen food, which he proceeded to unpack on the top of his desk.

"Celebration!" David announced, "We've got the Felixa job on our terms..."

Everyone in the room was delighted. The margin on this

contract would help pay for the new Wesdrill and would guarantee substantial year-end dividends to all the employee/shareholders of the firm.

"...and we start mobilization immediately."

Everyone groaned.

"Or, at least as soon as we've all had something to eat. I am sorry to keep you once again, but with the weekend coming up we have to get as much done as possible in the next twenty-four hours."

With that, he began to fill his plate and accept a cold beer from Ken on his return from the kitchen. Easy banter flowed back and forth across the room while everyone filled their plates with delicious foods from a Robson Street deli-restaurant. When everyone was well-fed and had settled back in their seats, David switched on the cassette recorder on his desk and began:

"And now to work!"

*   *

Four hours later, David rubbed his eyes with the backs of his hands and sat back in his chair. The others in the room stood to stretch and yawn. One by one they went out to the can or into their own offices to make phone calls home. Ruth brought the coffee machine into David's office and when everyone had returned and been served the fresh brewed coffee, David started the tape machine once again and summarized the evening's proceedings.

"Let's run through this all once more to make sure we're coordinated. Ed?"

"Right, we assemble all the hardware and drilling

supplies—including spare parts, break-down manuals and so on—here at the Vancouver shop. We'll have CP Air's cargo specialist in to supervise the weighing and crating. Then we'll ship it on a special flight to Hanover the weekend of May 5th. The Wesdrill with its trailer will be loaded there, hopefully on May 6th or 7th, and the plane will proceed directly to Sana to arrive on May 7th. That's two weeks from this coming Tuesday."

"Ruth?"

"From this office, we line up a CAT and fuel and food supplies to be ready in Sana when the plane arrives that Tuesday. I'll make new travel arrangements for Norm and Jack to put them in Hanover in time to oversee the loading of the Wesdrill, just in case there are problems. We'll coordinate our buying in Felixa with Manning, Demarche etc.—and double-check that they have arranged accommodations and introductions for the drill crew. We check to see that each crew member holds a current passport with the necessary medical stamps and we make application for entry and exit visas and work permits through old Georges Tabora in Beirut."

"George?"

"Ken and I will contact the other members of the crew to inform them of the schedule and instruct them on how to prepare for the trip. We have the basic personal equipment lists that Pat prepared for the Salif crew and we'll make adjustments for the highlands climate. Ken will leave this coming Wednesday for Montreal and wait there for the other crew members to assemble. They will leave Friday the 26th for Rome and be in Beirut sometime Saturday to meet Pat. In the meantime, I will secure replacements for the crew members on their present jobs and square everything with their field supervisors as we have discussed."

"Georgina?"

"Whew! First of all, I'll arrange a separate set of books to run this mobilization as top priority. I'll visit Bert at the bank tomorrow and discuss emergency financing in case we need it, and make arrangements for another foreign exchange account and mail drop with American Express in Sana. I'll have Shirley prepare a set of statements to distribute to the shareholders explaining the project and the extraordinary financial moves we have to make over the next few months. I will personally be in touch with all of our suppliers to expedite the delivery of supplies to Ed at the shop, and I'll have to meet with the union reps to okay the bonus clauses and negotiate the overseas allowances for the crew."

"Excellent. Now everybody call in around eleven in the morning and four in the afternoon every day to give Ruth a progress report. If you run into any snags, I'll be staying here in the office for most of the next week. Good luck!"

II

Ferdinand Patrick "Pat" Delise stood alone in the hills of Lebanon with his Nikkormat camera held loosely in one hand and a glove-leather travel bag slung over the other shoulder. He stood spellbound amidst the ancient ruins of Baalbec, about 35 miles east and north of Beirut, looking up at the sleek columns of the Temple of Jupiter. Sixty-five feet high, the Corinthian columns blossom into life at their tops where finely wrought acanthus leaves support an entablature of flowers and eggs. Six columns remain standing within the ruins, the other four dozen are gone—scattered to the four winds with only bits and pieces left lying about on the ground. Eight columns were removed in the fifth century of the modern era when Justinian took them to Constantinople to stand in the apse of the Hagia Sophia. Others were broken down and carted off to various other sites.

The temple grounds are older by far than Jupiter, predating the Roman occupation by centuries. An ancient name— Three Stoned—refers to the original temple that was built upon a base of three massive blocks of stone cut from a nearby quarry and chiselled smooth. Another block, from later times, still sits in the quarry partially chiselled. Sixty-nine feet long and sixteen feet wide, it weighs more than a thousand tons and is the single largest piece of building material ever manipulated by mankind.

Baalbec is very, very old. The Arabs tell in their stories that it was built by Cain to shelter him from God's wrath after his murder of Abel. Others say it was built by old King Solomon in the 9th Century B.C. Corrupted by the beliefs of his foreign wives, he gave himself to the ancient goddess Astarte, building this city in her name. More ancient peoples kept silent about its origins, implying that it had been here forever.

From the very earliest times, there was a certain stigma of

musty evil which hung about the place. The western cultures know the city by its Greek name, Heliopolis, and it was here that Jupiter and Astarte and far more primitive archetypes were worshipped. The customs of Heliopolis caused the place to be admired by some peoples and vilified by others. Women were considered common property in reverence to the goddess of fertility and so, ancestral bloodlines became untraceable from the very first generation. Here one of the greatest displays of art and architecture of ancient times served as a backdrop for human sacrifice.

In early times, the Hebrews preached against the excesses here and the city became something of a pariah to those for whom monotheism became the basis of belief. The Greeks were repelled by much of what went on there. But when the Romans came, they simply changed the primitive rites to accommodate Roman deities and resurrected many of the old practices. They built the fabulous structures that now stand in ruins. That which remains preserves the Roman artistic genius of the first three centuries A.D. in finer detail than any other monuments left to the modern world. Septimus Severus minted a famous coin set to commemorate the completion of this Temple of Jupiter. Finally, the Christians from the Eastern Empire under Theodosius shut down the temples and left the place in ruins. But here in the vale of Beqaa and in the surrounding hills, the Roman settlements far outlived the fall of the Imperial Empire.

Pat Delise walked through the ruins, stopping now and then to raise his camera, compose an image and take a photograph. He wandered back and forth searching for that one elusive view that would reveal all the hidden meanings of the time and place. Climbing the ancient stone steps he changed lenses to take close-ups of the sculptures and mosaics. Perched up on the stone rubble, he framed panoramas of the grounds. All the rolls of film were carefully marked and coded in sequence. They would all be returned to Canada on the next flight out of Beirut and sent to his wife, Adele.

It was part of their very successful marriage—she followed him where she could and when that was not possible, he traveled for them both, keeping her informed of his adventures through the pictures and long amusing letters. So, this day belonged to Adele. It was Pat's third day in Beirut and he knew it would be the last day for awhile that his time was his own.

Visiting Baalbec was a nice change from supervising the drilling of the salt dome at Salif. He had come out to Beirut to meet the new crew heading into Felixa and to help them organize their papers. Pat knew that it would take another two days before the company's contact here could complete arrangements for the crew's entrance and exit visas and work passes. The wheeling and dealing involved in obtaining such papers was bound by a complex protocol understood only by those dealing constantly within the hierarchies of the various Arab states. Pat knew enough to steer the arrangements in the right direction and then step aside and find other pursuits to occupy his time.

The other members of his highland crew would arrive tomorrow from Rome and they would need Pat's special guided tour of Beirut and its casinos. That would take two days and by then the papers would be ready. With any luck...

He considered himself fortunate to have had a day of socializing with the American crew from the oil rigs drilling on the Red Sea near Triangle Island, just north of Salif. The Americans had arrived the day before and they were quartered on the same floor of the hotel where COREX had reserved a suite of rooms. They had been flown out to Beirut for their R&R 'rest and rehabilitation', and today the Americans were getting an early start on the casinos where Pat would join them later tonight. First, he wanted to capture Baalbec for Adele.

*   *

On the first day of May, David Dixon strolled out from his hotel through the streets of Beirut. He was going to check out another casino in the international district. Last evening's casino winnings sat easy in his pocket. A modest amount to be sure, but special because it represented winning instead of losing. And for a variety of reasons that was an important feeling to David on this day.

The spring evening was gorgeous and yet, for David, it held a sense of foreboding. He was uneasy, alternately impatient and depressed, and his mood certainly had nothing to do with the weather. This should be a fun time, he thought, the first opportunity that I've had to relax in weeks. But it isn't. Why? That was the question that disturbed him most.

He was very conscious of the armed men who patrolled the streets of the hotel district. They were always here, but they were becoming more prominent in the scheme of things as the country grappled with the Palestinian question. David felt that things were coming to a head in the political arena and then... who knows? Many people here were making quiet preparations to leave Lebanon. David wondered again tonight whether this might be his last visit here.

Pat's drill crew, which left Beirut for Hodeidah just five hours before his arrival yesterday, seemed to have had the time of their lives here. COREX had rented a suite of rooms for the week and David had found notes of greeting and messages to him on the coffee table. Pat and Ken and Maurice and the others had left behind suggestions for his evening's entertainment. They left behind a half-bottle of Scotch as well, which he appreciated.

Some hours later, David was moving around the floor of the casino going from game to game, table to table, group to group, when he was hailed by an American voice. It belonged to an oil-man, a driller off one of the rigs working at Triangle Island just north of Salif. The man was part of a group of

Americans and their women and David wondered how they knew him by name. He allowed them to steer him clear of the roulette table and accepted their offer of a drink.

The women with them were professionally beautiful, also Americans, but David was quite sure they were working the hotels here in Beirut—and not the rigs off Triangle Island.

Finally, the man who first called to him took him aside and slipped him an envelope with more than $1,000 in cash inside. David questioned this windfall.

"You'll be in Salif before me, Mister Dixon, and I'd appreciate it if you'd give this here envelope to Pat Delise," said the driller, "Just before he left he advanced me a small sum to gamble in his honor. And this is the split of the winnings."

"Whew!" David whistled, "You must have had quite a streak of luck!"

"Luck? That's not luck does that to money in a place like this—that's talent! Ask Pat, he knows. We used to spell one another off at the tables, in days that are long gone by, or did you not know that he had that talent, too?"

The man smiled and David tucked the envelope into an inside pocket of his jacket.

"I will very definitely give this to Pat the moment I see him."

The hour was late and the crowd was beginning to thin out, leaving for other bars, casinos, restaurants, going home or to other unnamed places. It was typical of the weekday nights in Beirut in spring. David and the American's party chose to accompany one another back to the hotel. As they rode up to their floor in the elevator, the oilmen mentioned that they were expecting more female company to arrive soon at their rooms. They invited David to stop by for a drink, but he gracefully declined their offer and suggested instead that they all meet for breakfast or lunch the following day. This suited the general mood of the gathering and they parted in the hall with 'good night's'.

When David entered his room, he felt a sudden urge to talk to Margaret. It took a while to place the call to Vancouver, so he made himself comfortable and lit up a cigar. When Margaret answered the telephone, she sounded strange.

"David, oh David..."

"Hi, love..."

"Oh David!..." and then she was crying.

"Maggie, what's the matter," David was alarmed, sitting here helpless to comfort her half a world away.

"Oh David, Henry Svenson is dead!"

"Good god..." David was stunned. Henry was his best friend, almost a brother. They had grown up together on the coast, went to the University of British Columbia for their early training and then went their separate ways for the years it took them to develop their professional interests. Back together in Vancouver in their late twenties, they both took jobs with the same outfit, one of Canada's mining giants. David was a graduate mining engineer, Henry a geological engineer. When David moved over to the old McKinley Diamond Drilling Company as assistant manager, Henry went off on his own to set up shop as a consultant. When David and Pat and George and Norm and the others decided to buy the firm away from old man McKinley, Henry helped them arrange valuable industry contacts and financing. In later years, COREX was a top contractor for Henry's firm: CompLex Ltd. His death was totally unexpected.

He was one of the top exploration consultants in North America, and he would be sorely missed by the mining fraternity and the oilmen as well. David was stunned: they'd had a date to go fishing up north, next month.

"How did it happen?" he asked Margaret.

"It was a helicopter crash...in the Arctic Islands. He was supervising an oil survey for the government. They were up near Baffin Island when the machine just dropped out of the sky, oh David! They all died—all four of them...but... Henry?!"

David tried to comfort his wife over the phone, but there was not much to be said. This loss stung them deeply, both of them, but they had few words for one another. Tomorrow there would be time for the consolations, the arrangements to be made—now there was only shock and their immediate grief and that was so hard to share over a distance of thousands of miles. They talked for a short time together and then ended the call promising to be in touch, one with the other, in twenty hours time. David got up to turn off the lights, poured himself a Scotch and sat down again to stare out into the night, past the glittering lights of the Beirut waterfront.

III

The new drill crew arrived in Hodeidah late in the day. They left Beirut early on the morning of April 30th, but they traveled to Felixa by way of a most circuitous route. Having stopped at Damascus, Cairo and Aden before crossing the Felixan frontier, it was late afternoon before the five stepped down from the plane to the desert at Hodeidah.

Pat Delise was glad to be back near the work. It was not that he would ever mind a holiday or a chance to travel, it was just that he accepted a certain responsibility for the work and preferred to be on a job site until the coring was completed. The other members of the crew were excited to be in Arabia for the first time.

During their time enroute from Montreal, the four new drillers formed personal alliances that spoke well for David's choice of personnel for the job. Ken Holby became good friends with a younger driller, James "Smitty" Smith, who worked for COREX in eastern Canada—in northern Ontario and the Maritimes. Maurice Leblanc, the other senior driller, worked out of Montreal as well, drilling around Hudson Bay in the north and in the urban Montreal-Quebec City areas. A natural partnership formed between Maurice and the other young driller, Tom St. Martin, an Indian from the Prairies who worked the western region. Tom had once worked a job in northern B.C. with Ken Holby and considered him a good friend as well. It was Ken who had recruited Tom for Felixa. Pat was interested to see how these partnerships worked and allowed human nature to arrange his crew staffing for him. Ken and Maurice were the obvious leaders and it was curious to see the east/west pairing occur quite naturally in each case.

When the men cleared customs they found Billy MacDonald and Mike Perrone waiting to greet them in the airport lounge. That was unusual, one of them should have been at work unless there was a major breakdown. But no, thought Pat, something more is going on here. It was Billy who took him aside to explain:

"Pat, we've got a real hassle here. Gordy took off and we don't know where he's gone."

Billy went on to explain to Pat that Gordy had left them under strange circumstances and that the two younger drillers had called in the security officers from Hodeidah thinking he might try to leave the country from the airport here. He was still nowhere to be found and certain Felixan officials were beginning to growl about that.

Pat remembered how he had taken both Billy and Mike aside when he heard from Vancouver about the breakup of Gordy's marriage to Marie. All three of them had been watchful, hoping that they could act as a balance for Gordy if he felt the need to react. Gordy was in obvious distress and they stayed close to him and helped smooth out his anger and frustration. All was normal again before Pat had left for Beirut. For some reason, he had not maintained his balance for very long.

"Mike and Gordy came into Hodeidah yesterday to do some touring and to pick up our mail at American Express," Billy continued, "Gordy went over for the mail while Mike stopped in to see the expediter about another change of menu. Gord was supposed to meet Mike in an hour's time at the jeep, but he never showed. After a bit of a wait, Mike and a couple of locals from the import house went looking for him. No luck. When it began to get dark, Mike had Salim Bin Shazar drive back out to the mine to get me.

"That's when we ran into problems with the security people. There is that damn curfew for foreigners and we couldn't be out on the streets. You know they still hold to the Muslim laws here: an eye for an eye and all that. ...We didn't know what kind of situation Gordy might be getting into so we went to see the local security chief. We tried to explain what had happened, why Gordy was upset and what he might do, but they just didn't seem to understand. They kept *us* there, told us we couldn't be out after curfew.

"A little while later, an officer showed up with a handful of mail addressed to us that had been returned to American Express by a young Arab. Only the Express office wasn't open, so he ended up giving them to the police to guard. He had been paid to take them back to the Express office by a foreigner down on the docks.

"That really got us worried—we thought Gordy got kidnapped or something, until the boy described the man who gave him the letters. It was Gordy and he sounded like he was very happy. He must have been celebrating! The police just wouldn't or couldn't understand such a thing and they wanted to call in someone from the national headquarters at Sana—top priority, an international incident. We cooled them down a bit, but they are getting very upset with us. They want to see you right away, boss. They're mad..."

Pat took a deep breath. Jesus, he thought, just what I need! He would have to go see the police right away. He called Mike over and asked him to take the new crew to Salif and get them settled in. Then he left with Billy for the security station.

When Pat emerged from the building hours later, he was trembling himself with anger and frustration. Billy was waiting for him in the lobby and quietly followed him outside. Billy had been through a very frustrating time with the police because he didn't speak the language. Pat was relatively fluent with basic Arabic—his frustration derived more from the attitude of the people he talked with than from language difficulties. A typical bureaucracy, he decided, ignoring the simple human problem and looking for more exotic implications. He had tried to explain the possibility that Gordy had gone underground looking for a bottle of booze,

49

but the security people refused to acknowledge the logic of liquor. They dismissed this as a concern, arguing that there was no liquor permitted in Muslim countries and denying the existence of the substantial blackmarket in booze that flourished wherever the law forbid its presence. The interview finally ended in a stalemate—just short of open warfare. Pat held to his conviction that Gordy was simply looking to blow off steam, the police suspecting a plot with international implications.

It was beginning to get dark once again and there was no time to arrange transportation back to Salif. Pat and Billy took a room at one of the old hotels downtown. It was extremely hot and the hotel was very old.

The center of Hodeidah was a classical Arab city, surrounded by miles of tin shacks and mud huts. There was some running water, but even that had to be boiled before consumption. The outlying sections of town got their water supplies from tank trucks that would park in the neighborhood marketplace. With the markets closed for the night, the city was quiet, the silence broken only by the call from the minaret of the mosque. Pat and Billy lay down to get some sleep.

It was after midnight when they were awakened by security officers and driven back to see the chief. The chief was very reserved and treated Pat with disdain. He explained that his men had located Gordy down on the waterfront and were holding him there. He invited the Canadians to accompany him to the spot.

They were driven to the docks as part of a small procession of jeeps and found a very exuberant Gordy Watson surrounded by searchers. He had evidently found a Greek ship captain in the city with a similar inclination to celebrate that previous evening and they had gone back to the ship to do so. Nearly twenty-four hours later, the ship captain heard that the police were after Gordy and dumped him ashore just before leaving port. Gordy woke up after dark and went on a bit of a rampage on the dock as part of his way of venting his personal frustrations. And so it was necessary, explained the chief to Pat, to take him into custody until the damage was assessed and the courts decided what to do about his drunken escapades.

Pat asked the chief to send for the expediter and the agent for American Express and then began to negotiate Gordy's release. Some hours later, the chief agreed that it would be best for the entire matter to be settled quietly and as soon as possible. He agreed that Gordy's drunkeness was more a personal tragedy than a flaunting of Islamic law—considering the family circumstances. The American Express agent helped the police assess the damages and vouched for the company's integrity. During the long proceedings, Gordy's mood changed from exuberance to sullenness, and Pat realized that this night was to be another turning point in Gordy's life.

Just after midnight, the chief of police allowed Pat and Billy to take Gordy back to the hotel after reaching agreement that he should leave the country as soon as possible. Gordy was quiet, no trouble at all. He had withdrawn inside himself and refused to communicate with anybody on the outside. Pat found this very disturbing, but he was relieved that the affair was over. Thank God, he though, that Gordy had not had a sudden craving for an Arab woman—that would have meant a nasty death for him.

In the morning, Pat arranged a ticket for Gordy on the afternoon flight to Cairo. Billy went back to Salif with a fast driver to pack Gordy's belongings and send the luggage to the airport. Pat took Gordy to the airport at noon to wait for his flight.

*   *

The Aeroflot charter flight from Damascus to Hodeidah thundered south over the Saudi Arabian desert. David Dixon relaxed in his window seat as the jet engines of the Russian Y-62 streaked silver contrails across the brilliant blue sky. The Ilyushin-62 was quite a nice plane, about the size of the DC-9, and this flight crew was very friendly. An indication perhaps of how hard the Russians were trying to establish a neighborly presence in the region. The flight was almost full, with a mixed compliment of passengers of many nationalities—all of them headed for the southern reaches of the peninsula. This charter flight was the most direct form of transportation to David's destination in Felixa, but he could have taken any number of other carriers to the southern desert with connecting flights to various points in Felixa. And that field of choice, he mused, was perhaps the most dramatic indication that the region was beginning to come of age commercially.

Or rather, come of age *again*, David remembered as he scanned the desert landscape below. Almost thirty centuries before, give or take fifty years, this region was active under the aggressive sponsorship of old King Solomon, David's son. In the course of building one of the world's great trading empires, Solomon salted colonies of his countrymen throughout the Middle East to encourage the native peoples to become commercially productive. Solomon developed any number of commercial enterprises on the Arabian peninsula, including a copper smelter at Ezion-geber on the Red Sea. It was partially to encourage Solomon to invest in her kingdom of Saba (Sheba), which included much of modern-day Felixa, that the Queen of Sheba traveled to see him at Jerusalem. Her people became part of Solomon's schemes and flourished for many centuries. They went on to colonize Ethiopia—just a short distance across the Red Sea, and be colonized in turn by the Christian Ethiopians themselves in the 6th Century A.D.

Less than one hundred years later, the forces of Islam swept across the Arabian desert to lay the seeds of the modern nations which rimmed the southern tip of the peninsula. It was only in the present 20th Century that most of the little colonies of Solomon's Jews left the area to return to their ancestral homeland in the modern state of Israel.

Watching from his seat just aft of the wing, David caught sight of his landmark: King Solomon's fabled gold mine, which bloomed on the desert between Medina and Mecca. Called the Mahd adh Dhahab, 'Cradle of Gold', the mine yielded perhaps 31 metric tons of gold to the ancients—most of it in the form of nuggets, wires and crystals near the surface that could easily be separated by simple panning and winnowing. Passing overhead, the passengers could see the great outcropping which contained the rich gold veins surrounded by clusters of mud huts, some dating back nearly 3,000 years.

The ancients left behind more than a million tons of mine dumps and waste rock which still contained a minimum 0.6 ounces of gold per ton, according to modern assay reports. At one end of the curved pit stood the modern processing sheds built by Saudi Arabian and Canadian interests when they reactivated the mine in 1939. They followed the veins underground throughout the Second World War until they were forced to close down when the economics of the operation became unwieldy. The flotation and cyaniding mill appeared silver against the gray-brown of the desert, but the mine tailings dump was a brilliant green and gray there on the desert. David knew that a team of American diamond drillers had been at Mahd adh Dhahab within the past year and that their shallow explorations confirmed enough potential, in the form of gold reserves, to reopen the mine once more using modern processing methods.

David was alerted to their approach to Felixa nearly two hours later when he spied the oil rig derricks near Triangle Island. The rigs were no more than ten miles north of Salif, drilling exploratory wells where the presence of the salt dome gave the petroleum people reasonably good odds for discovery

53

of useable reserves. The collection of vessels surrounding the rigs confirmed the American driller's boast that the whole operation was afloat, with ships serving as workshops and warehouses and dormatories for the men. The floating city was serviced by helicopter and small barges and lighters.

With any kind of luck, thought David, I can be out there at Salif by nightfall. He fastened his seatbelt, shifted position for the descent and turned back to the window. There was no way for him to know that Pat Delise waited only a few thousand feet below him, pacing the lounge of the airport, anxious for the plane to land.

Gordy Watson was there at the airport as well, with all his luggage packed and tagged and ready for his flight out of the country. He sat apart, in the corner, looking at nothing, observing nobody. Pat had tried all night to talk with him and to make him understand that he had the sympathy of the crew. A rest was all he needed, a change of scenery and the opportunity to roam around a city where the people all spoke his own language. Pat grieved with him, because Gordy was a fine man and a good diamond driller. But Gordy had shut himself off from other people. He had shifted into neutral— he was neither sullen nor angry—and who knew what it would take to make him human again?

When David's plane landed, Pat began to settle down. The decisions he had made were irrevocable, but were they right? He shrugged. He would speak to David. Soon. Soon.

By the time that David cleared customs, the announcement was being made for the boarding of the day's flight to Cairo. David barely had time to shake hands with Pat before Gordy rose from his corner and walked out to the plane. Pat laid a hand on David's arm.

"We've had a casualty, David. He's going home. He's stunned—doesn't feel a thing. I've tried just about everything, but nothing reaches him. I gave him travel money and something extra in case he wants to stop somewhere and let loose a bit, but somehow I feel that he is beyond that now."

Pat explained briefly what had happened in the past two days: Gordy's disappearance, the problems with the Felixan officials and so on.

"David, we've lost a good man."

"We've lost two, Pat," said David quietly, and then told Pat about Henry Svenson's death. He also told Pat about his stay in Beirut and gave Pat the envelope from the American oilman. There was silence while the two Canadians, partners and friends, totalled their losses and adjusted reality to accommodate the changes. They soon left the airport to travel to Salif, forty miles north.

*   *

The temporary presence of the highlands drilling crew bouyed the spirits of the crew at Salif and within a day things began to return to normal. The Canadians were using one of a dozen houses built for the mine's workforce during its last period of operation. The house was open, there were no doors or windows—just arches for the sea breezes to blow through, and it was reasonably cool and shady for a desert residence. It had high ceilings and very few furnishings except for those that the drilling crew brought with them or purchased in Hodeidah.

The house was only one hundred feet above the high-tide line on the shore of the Red Sea and to swim, one had to wade out one hundred yards through the shallows to find water over waist-deep. On their off-shift, the men would swim with the porpoises that were forever playing off-shore.

The Arab cook was a pleasant individual who had been the target of numerous complaints the first weeks of drilling here for serving nothing but mutton stews. Pat made arrangements with an importer in Hodeidah to bring in a sufficient stock of western foods from Denmark and Sweden. Things had improved immediately and everyone lost the green tinge that had plagued them. The cook was delighted himself with the change in menu and went out of his way to serve them well.

He brought David a cup of hot coffee as soon as he entered the room for breakfast with Pat and Mike in the kitchen of the drillers' house. At six o'clock, Billy and his crew of Arab helpers came off shift. The table was cleared after the meal so that Billy and Mike and Pat could review the work of the past ten hours. The helpers greeted one another as familiars and drank more coffee. They were all relatives of the mine manager, as were all the other 500 residents of the village. That was the pattern in most of the small villages along the coast.

The two drivers also came from the village. The driver of the Land Rover would bring two helpers from the village to the house to meet their driller, then pick up the other crew from the drill site at the end of the shift. After an hour or two of breakfast or supper, depending on the time of the shift change, he would drive the new crew out to the drill site, pick up the tank truck driver there and pick up the others at the house to take them all back to the village. It was an arrangement that worked well. The two drivers each spoke a bit of English and could talk with Pat in Arabic to some extent as well. They acted as interpreters for the COREX crew when that was necessary. The drill helpers soon learned the work routines and there was little need to talk about technical problems. The drillers themselves were fast learning to make themselves understood in Arabic.

The tank truck driver ran a slightly different schedule and David had yet to meet him. Right now, he would be down at the beach near the village two miles away pumping 1,000 gallons of salt water into the tank mounted on an old truck chassis. The COREX crew had brought a gas-engine pump with them, so it was not strenuous work. He would drive up to the drill site at the time of the shift change and leave the truck there with the next shift's drill water supply, before catching a ride back to the village with the other vehicle.

At eight o'clock, David went up to the drill site with Pat and Mike's crew to see how things were going. The mineral deposit here was in the form of a great salt dome rising from the edge of the sea. In the early days, the salt was mined from an open-pit by hand with long pointed sledges and transported in woven baskets. The salt was a primary trading staple of the Arabs. The caravans traveling the Incense Route through the interior of the Saudi Arabian peninsula depended on this source for trading supplies for thousands of years. In the 12th Century A.D., salt was mined in Africa for more than twice its weight in gold.

In later years, the salt was used as a foreign trade staple and was barged out to ships which stood off the shallow beaches. COREX was there under contract to Manning, Demarche & Barrett to determine the extent of the remaining deposits so the mine could be reactivated to provide Felixa with foreign currency and to provide jobs for the local residents.

When the crew reached the job site, they checked the oil and fuel supplies for the drill, tested the brine solution and went to work. The drill they were using was an older Boyles Brothers' BBS-1, which Ed McGavin had rebuilt at the Vancouver shop especially for this job. It was skid-mounted with a small tower hoist. The machine had been equipped with an hydraulic chuck and rigged with a larger head to accept 'N' series drill rods. With a more modern power supply, the 40 horsepower drill was capable of coring to depths of about 700 feet. To drill in salt, Pat Delise found that the only drilling fluid that worked was water that was supersaturated with salt.

Clear water used as drilling fluid was disastrous. It dissolved the sidewalls of the drill hole and corroded the core sample itself. But it could be used for other purposes. The casings used to guide and steady the first few yards of drill shaft were cemented in place near the top of the hole to prevent them from moving around through the softened salt. They were retrieved for reuse using fresh water to dissolve the salt around them. Pulled from the hole, the cement was broken off the steel casings with a sledgehammer.

The salt body was striated with bands of sandstone which dirtied the drilling fluid. The drillers had to change the water in the hole two or three times per shift, using almost the full thousand gallons of water in the tank truck. One helper was kept constantly busy mixing raw salt into the sea water until no more would dissolve.

The coring was going very well. After six weeks on the job, the crew at Salif had finished ten of the twenty-four holes they had contracted to drill. They drilled in ten-foot runs, pulled the core barrel to unload the sample, dropped an empty barrel in place and drilled again. One helper, under the driller's supervision, carefully unloaded the core samples and laid them down into the narrow channels of the core boxes, using small wooden markers to keep track of the footage. Averaging one run of ten-feet per hour, the crews were pulling one hundred feet of core per shift. None of the holes was much more than 600-feet deep. A layer of bright shale was found at the base of the salt dome and this was used as an indicator of the end of the hole.

The Land Rover was used to tow the skid-mounted drill from one hole location to another. The diamond loss on the job was negligible. Each bit remained sharp for more than 3,000 feet or nearly a half-dozen holes. Mike's crew was still using the second bit. The salt was very high-grade and easy to drill.

At the end of the shift, all of the equipment including the hand-tools and the wooden drilling platform were washed in fresh water to halt salt corrosion. The drill was still in magnificent shape, but the repair shop had already warned David that the machine would have to be torn completely apart after this job and rebuilt once again.

David and Pat returned to the house after about an hour at the mine. They discussed all the events of the preceding days and decided that Billy and Mike were quite capable of handling the rest of this job on their own. Pat wanted to stay at Salif for a few days more just to smooth over the strained relations of the security people in the area and to see that the work was not interrupted by the repercussions of Gordy's actions. Tom and Smitty, the two younger drillers, would stay with Pat so he could instruct them in the local customs. Ken and Maurice and David were to leave for Sana to meet the charter plane bringing the Wesdrill. Pat and David then went for a swim.

IV

Traveling the hundred-odd miles from Salif to Sana was like moving to a completely different world. David had seen Sana before, and so had Pat—when they visited the city last year in preparation for bidding the job at Salif, but he was still excited by the prospect of establishing a camp here in the highlands. For Ken and Maurice, this was their first look at this part of the Middle East.

They drove toward Sana on the country's only paved road, built as part of the renovation program of the port at Hodeidah with the aid and funding of the People's Republic of China. The gravel roads which connected Sana with the other highland cities were constructed under the sponsorship of the American foreign aid program. The Russians were modernizing the port at Ras Kadhib, and both the Germanys were developing light industry in various places around the country. David found it ironic that in the midst of all this foreign activity, Felixa was still looking for one trustworthy ally to help her preserve her own interests in this beautiful land.

As the Land Rover wound its way over the western foothills, the COREX drillers and their boss got a good look at the country. It was not hard to see how Felixa got its Roman name: "fortunate, fertile". The greenest, most productive land in Arabia spread out before them. The land made its magic presence physical as scores of different perfumes and scents rose off the fields of coffee and spice to tantalize the nose. The mountains and hills were terraced and the fields lovingly tended. The people who lived here were descendents of the strongest tribes to inhabit the southern Arabian peninsula in earlier times. Drawn to the rich soils of the highlands and the moderate climate, the strong tribes had displaced the weaker ones and taken over these heights. The weaker tribes were forced to move down to the Tihamah or to wander as nomads in the desert. A class system grew up in this way, finally pitting one sect of Islam against the other so that there was constant warfare throughout the land for hundreds of years. The highland peoples built fortress-like villages high up on the mountain slopes overlooking the fertile valleys and here they remained.

David recalled that Felixa was divided into four geographical sections, each with its own climate. Salif and its salt dome were part of the Tihamah, the coastal plain desert which bordered the Red Sea for three hundred miles. The Tihamah was roughly thirty miles wide and gave way in the east to a ridge of high mountains which ran the length of the nation. Beyond the mountains lay the great central plain, a plateau above 4,000-ft. elevation. The eastern slopes of the mountains fell away into foothills which divided the great central plain into large natural fortresses. It was here that most of the population of Felixa lived.

The eastern edge of the plateau was intersected by a great many *wadis*, dry streambeds and rivercourses which flowed only during the times of the great rains. The land appeared corrugated in its gradual descent to 3000-ft. elevation where it leveled off into the eastern desert, Rub' Al-Khali, the 'Empty Quarter'.

The mountains scraped the bottoms of the clouds drifting east from Africa and west off the Indian and Arabian oceans, which dropped almost 20 inches of rain on this land each year. The air was still a bit brisk, but David knew that in July the average temperature would be in the neighborhood of 70° F. That temperate weather would make the drilling work so much more pleasant than in the sweltering heat of the Tihamah at Salif.

And with any luck, they would find something here to justify the government's gamble and Harry Demarche's plans. The highlands had areas of very rich volcanic soil and heavy metals were sometimes found in relatively pure form in gas-bubble cavities of ancient lava beds. Also, most of the highlands were rifted shield areas of igneous intrusive granite and in areas where igneous intrusions occurred repeatedly, heavy metals were often a good bet.

The tallest mountain in southern Arabia, Jebel Hadur, rose 12,336 feet above the city of Sana and it loomed above the Canadians as their jeep drove into the town. Sana itself was an ancient walled city where almost everyone was on foot. They passed through the columned arch made of brown mud bricks that served as one of the gates to the city. The dirt streets were quite wide and lined with tall houses, almost like European row houses, that were made of stone or stucco or mud bricks. Four and five stories high, these buildings had openings for windows which were barred with ironwork in tracery of very delicate design. The Citadel and the Great Mosque dominated the skyline.

Nearly 125,000 people lived at Sana or in the vicinity and it looked like most of them were in town this Monday, May the 6th. The driver of the Land Rover had no trouble finding the American Express office, where he waited while David and Ken and Maurice went inside to introduce themselves and pick up their mail and chat with the English-speaking manager about local customs and so on. The men met with local officials, visited the airport to size up the CAT they had arranged to lease, and then went driving off into the countryside to find the job site and their new residence. The Wesdrill was scheduled to arrive the next day. There was time now to unpack and relax before the action got underway.

*   *

The morning of May 15th dawned crystal bright with just a bit of chill in the air. By ten o'clock, the world was filled with sunshine and warmth—so long as one kept out of the wind, which was how Pat Delise was enjoying this morning. He sat on the hillside above the drill set-up, watching Maurice and Tom work their third hour on shift. There was a small pocket in the hillside at right angles to the wind and he was quite comfortable. The men were working on their second drill hole in this area, just twenty miles east and north of Sana near the head of the Wadi Abrad.

The pilot hole had gone well and the first boxes of core had been shipped out by plane the previous day. They would be in the hands of Sam Ellis no later than Monday the 20th, and perhaps they would give the principals at Manning, Demarche & Barrett good cause for their optimism. Or then again, they might not...but that wasn't Pat's problem. He earned his money and lived his life making sure that the holes were drilled on schedule, on budget and with the greatest possible percentage of core recovery.

Still, he hoped that they would find something in the samples that they liked. It always helped to take some of the pressure off the drillers. And that was a boon to Pat—he could take a few minutes then to relax and reflect on things.

From where he sat he could look out across the valley and watch the people farming the slopes of the green terraced hills. These fields and vineyards and orchards produced all kinds of fruits and vegetables and coffees which were now part of Pat's daily diet. There were no automobiles here except for their own Land Rover. Everyone else used camels or donkeys as transportation. Pat watched the camels wind their way up the hills to the villages which were jumbles of buildings stacked on the flat tops of the mesa-like hills.

Every so often he would point his camera here or there to take a picture. Many of his pictures included the drill rig, because it was so prominent in his line of vision. The drill had arrived on schedule, all eight thousand pounds in two crates. The support gear weighed in at another five tons. It took two days to truck all of it up to the job site, but only twenty-four hours to reassemble the machine and put it to work. When David finally left Sana to fly to Cairo for his connection to London and then home, he took with him Pat's pictures of the highlands for Adele, plus a couple of rolls of film for the Wesdrill people in Richmond—showing their machine at work on this mountainside in southern Arabia. It was a small gesture, but one that helped COREX maintain its reputation throughout the industry for service and courtesy in the highly competitive world of exploration drilling.

Thinking back, Pat realized that it was just six days before that David finally decided to leave. He stayed long enough to see the drill in place and working to satisfy himself that he had made all the right choices in equipment and manpower for this job. When Pat saw him to his plane at Sana, David's face was beginning to show the strain of two weeks constant travel, the problem of Gordy Watson and Henry Svenson's death. The man needed a rest. Badly...

Pat frowned, thinking about David and what his energy and business skills meant to COREX. We must be careful, he thought, to see that he takes care of himself. Perhaps Norm and George and I should take on more of the responsibility for things...and these thoughts made Pat feel a bit guilty about lying around in the beautiful May sunlight. He shrugged off this feeling when he heard Maurice call him from below.

He looked to see Maurice waving him down. There was a problem of some sort for him to solve. Well, he thought, that's how I earn my money...

"It looks like we've hit a fracture, Pat," Maurice told him as he showed his drilling superintendent the sudden change in the graphic analysis of drilling pressures in the hole. "And we're losing a good bit of water."

The men looked at the print-out from the graph and checked the readings of the other instruments built onto the

Wesdrill to determine what had caused the variation.

"It looks that way," replied Pat, "well, better pull the core and take a look."

Maurice nodded assent and signalled for Tom to stop the machine and fish the core barrel out of the drill hole. Tom introduced the wire line recovery coupling into the drill shaft and dropped it down to connect with the coupling on the upper end of the core barrel. Using the wire line hoist on the Wesdrill's tower, he slowly pulled the core barrel up the inside of the string of drill rods. When it reached the surface, Tom very gently lifted it clear of the shaft and carried the five-foot barrel away from the machine. He returned with an empty barrel which was soon lowered into place and coupled to the drill shaft just behind the diamond head.

Using a special pair of wrenches, Maurice pulled the core barrel from its lift-out casing, then pumped the inner barrel free with water under pressure. Gently cracking the split-barrel open, he laid bare the core sample itself. Tom and Pat came over to look at the core which he transferred to its snug channel in the core box.

"Looks like a fracture all right, and it appears we might be in a bit of friable ground here—with this stuff spilling out here..." said Tom, sifting through some of the crumbled rock in the core box. It was this crumbling rock that would prove a problem, falling down the shaft and jamming behind the drill head if it were allowed to vibrate much longer.

"I think maybe we'll cement it then," said Maurice, "What do you think, Pat?"

"Yes, that would be smart until we get to know this ground better. You just reloaded the core barrel, did you Tom?—well no matter, we'll be pulling the whole thing now anyway..."

The drillers switched the machine out of neutral to begin extracting the drill rods from the hole. The 100-horsepower drill operated with a heavy-duty synchro-mesh transmission and hydraulic clutch very similar to that of a truck. This allowed them to hoist out the heavy string of drill rods stopping every couple of yards to uncouple the rods one at a time, without jerking the whole string violently up and down in the hole.

When the string was free, the drill head was removed, the diamond bit taken off, and the string of hollow rods dropped back down the hole. The column of water, which circulated constantly in the drill hole to cool the bit and carry the debris to the surface, was pumped out as much as was possible. A fast-drying cement was poured down the hole displacing whatever water was left at the bottom of the shaft to fill the fracture. The cement was left to set until the next shift, and Tom and Maurice cleaned up the rig and filled out their reports. Pat walked down into the village to locate their driver and arrange for him to retrieve the drillers and the core boxes immediately. Pat himself set out on foot to cross the valley to their house. He was hungry, it was just past noon.

Eight hours later, he rode back out to the rig with Ken and Smitty to begin the night shift. They put down a string of rods with a new diamond bit and proceeded to drill through the cement which was holding the broken ground in place. When the instruments showed that they were at the bottom of the previous hole, about to re-enter the rock, they pulled the core barrel to discard the cement plug. Then they continued their work.

"How goes it?" Pat asked, "What do you think of this rock?"

"It's as hard as the hinges of hell!" snorted Ken, "So I guess it should be smooth sailing from here on out..."

The work continued at a steady rate over the next weeks

and the men settled into their new lifestyles with very few problems. Each week, Pat left for a day and went down to Salif to see how Billy and Mike were getting along. Occasionally there would be visitors to the drill site. Otherwise, it was ten hours on shift, ten hours off shift, with two periods of shift change at breakfast and dinner when the five Canadians were together for a short time. It was pleasant, working in these mountains in early summer.

*   *

On Tuesday, May the 28th, Pat watched a different Land Rover careen across the landscape, banging and bumping its way up the slope to the drill site. As it came closer, he saw that the jeep was filled with people and then he realized that he knew them all. The vehicle came to a screeching halt not twenty feet from the Wesdrill. Harry Demarche, David Dixon, Norm Harrington and Jack Chong piled out from the back while big Sam Ellis unfolded his long legs from the front. Muhammad Bin-Alfir, their driver, was grinning from ear to ear. It was obvious to Pat that one of these tall northern strangers, speaking with a fair degree of authority, must have assured him that Land Rovers were *meant* to be driven fast up mountains—so they didn't get stuck anywhere. ...The ruse worked, he was only too happy to oblige and, once he conquered his panic, the trip from Sana had been a real thrill. It took a few minutes for everybody to catch their breath and to renew introductions.

"So what brings all of you out here...back here...so soon?" Pat asked them.

"I liked your colors," said Sam Ellis.

"It looks like you've already found something here that we can work with," added Harry.

"Well, we were hanging around London with nothing to do, so..." began Norm.

"I decided I needed a vacation!" Jack Chang groaned, and everybody laughed.

"Actually," interrupted David Dixon, "we all flew in for the party."

"Party?"

"Yes, tonight at the British Embassy in Sana. Close down after this shift. Everybody goes with us. After all, it's in your honor!" David chuckled, looking around at the drillers, who looked confused, scratched their heads and went back to work. David gave Pat a touch on the shoulder.

"We're going up to the house to tell Maurice and Tom. We'll meet you there as soon as you can shut down."

"David, how long have you guys been back?"

"Oh, a couple of days..."

"You must have spent all of that time arranging this."

"We've been working like hell! But this reception is just a spin-off from other things. A chance to meet some of our neighbors. We'll talk up at the house. You can bring Harry and Sam up to date and so on before we leave for town. See you at the house."

And then they were gone again, blasting down the mountainside in a cloud of dust at breakneck speed amid the joyous whoops of Muhammad Bin-Alfir.

*   *

With Muhammad leading the way, it took less than an hour to drive the twenty miles into Sana. The reception was being held at the home of the British ambassador, acting on behalf of the Canadian government, rather than at the embassy building in the heart of the city. The Land Rovers were not altogether out of place among the other vehicles parked on the slope leading to the house. There were, to be sure, a few classic official cars, but there were also a number of older vehicles including three more Land Rovers in various stages of ornamentation.

The men presented their hand-written invitations to an Arab in ceremonial dress who stood by the door. The man wore a white robe cinched with a leather belt that was embroidered with silver thread. His brightly-colored turban was wrapped over a skullcap with silver embroidery. He had his ceremonial *jambiya*, the curved dagger, thrust beneath his belt.

The house itself was typical of the homes of the wealthier highlanders of Felixa. The lower part was built of blocks of granite and the upper stories of mud-brick. The house perched on the mountainside facing east, with the entrance at the top of the slope. The upper story contained the formal hall which opened out onto a stone parquet terrace or loggia. The lower stories all opened out onto balconies or terraces as well.

Inside the door, the Canadians were greeted by an aide who introduced them briefly to the ambassador and the governor of eastern Felixa. As a group they were advanced along an informal reception line to greet the trade consuls of various countries involved in technical aid projects in Felixa. The East German representative described the electrical power station that his countrymen were constructing down in the city. The Soviet consul talked of the work of his people in modernizing the port of Ras Kadhib. A couple from West Germany greeted them with great warmth and talked of home. The man inquired about the Wesdrill that had arrived from Hanover and promised to visit the drill site to see the machine in operation.

When they reached the end of the line, the Canadians conferred briefly with the American oil people present who had just gotten details of the scoring of the previous day's Stanley Cup hocky game. Within the hour, everyone was circulating freely throughout the hall, drinking a variety of coffees and teas and fruit drinks, and sampling the food that was being set out for the guests. The foreigners present discreetly slipped outside to the balconies to smoke their cigars and cigarettes.

The ambassador moved from group to group with his wife, making all the guests feel at home in his house. The governor also moved slowly among the guests, stopping to spend a few moments with each and every person individually. He would speak through his interpreter, asking everybody about their homelands. Then he would ask: "And what do you think of my country?" The interpreter was handling close to one dozen languages within the space of half an hour—a formidable accomplishment.

Harry and David were talking together near the arch opening onto the loggia when the governor approached them. He was an impressive figure in his turban and robes. Harry was about to slip quietly away so that the governor could speak privately with David, but the governor gestured that he should stay and talk with them. He questioned both men about British Columbia and the Canadian northland. Then, without asking their opinion of Felixa, he said:

"My colleague the governor of western Felixa tells me that you have drills that can penetrate any material to any depth. He is very interested in locating new sources of water for his people who barely survive on the Tihamah. When you are finished drilling at Salif, perhaps you could drill for him?"

David began to explain the technology of exploration drilling to show the governor that it was not exactly suited for well drilling. The official frowned:

"But could you not do something for my people in that desert?"

David looked to Harry, who shrugged and said:

"I suppose we could bring in some experts and the right equipment..."

"Good!"

The governor clapped his hands together in response, beaming a smile and continued:

"Now tell me of your present work here in the highlands. I trust we shall be good neighbors for some time..."

David briefly described the work that Pat and the others were doing and invited the governor to come for a guided tour. It was immediately arranged for the next day. Then the governor took Harry's arm and indicated a quiet corner:

"Let us talk of your mine..."

*   *

Pat left the party inside and walked out on the loggia of the ambassador's residence. He stood in the shadows looking out over the foothills and wadis which stepped gradually down to the desert in the east. Footsteps approached him from behind, sounding hollow on the stone floor of the terrace. David came up alongside to lean on the parapet. Together they watched moonbeams dance across the landscape.

"Imagine what it must have been like two...three...

thousand years ago," Pat marvelled, "when this was the kingdom of Sheba. This house wasn't here, all the action was over there: at Marib, eighty miles or so east in the desert. The legendary Marib. That's where the Queen of Sheba had her capital, and what a place it must have been! All the traffic on the Incense Route stopped at Marib to rest and resupply and trade. It was an oasis for the spice caravans coming out of the Empty Quarter, the Rub' Al-Khali, and Marib funneled all the good things grown here in these highlands right into those same caravans for trade throughout the Middle East and North Africa. Because of Marib, Sheba was the richest of all the Arab lands.

"It was such a beautiful place to stop after a desert journey. Sheba's priest-kings, the Queen's heirs, built a massive dam across the lower end of the Wadi Abrad in the 6th Century B.C. Two thousand feet long, the dam was the keystone in a grand scheme to irrigate an enormous section of the southern Arabian peninsula," Pat gestured, sweeping his arm across the panorama of terraced hills in the night, "And it worked! For twelve hundred years the dam held back a vast lake of fresh water. And then it burst, in the 6th Century A.D. It was such a spectacular event that its destruction is recorded in the Koran. We'll go there this week, maybe—we'll go out and see the ruins. There is still part of a magnificent 5th Century temple in ruins, all marble..."

"Maybe," said David, "yes, that would be something to see."

After a brief silence, Pat turned to David to appraise his mood.

"You're quiet tonight, what's up?"

"Oh, just thinking," David responded quietly. "Listening to you talk about the Queen of Sheba reminded me—today's my Margaret's birthday..."

"Did you leave her a present when you were home?"

"Well, I never did get home, you see," David told him, "Norm and Jack and I got together in London and then things began to move. Harry got ahold of us and told us to stay put, he was heading out this way within the week. Plus, there were the preparations to make for the Malaysian job..."

Pat looked confused.

"...so we stayed in London. Margaret came out from Vancouver last week. We had a good time."

Pat sighed, thinking of Adele. David lit one of his petite cigars and handed another across to Pat.

"Oh! And I talked to Adele," David continued, "when I called in to the office just before I left London. She'd come in to collect the latest batch of film. She said to tell you she'd dragged out her old skirts and veils from that time in Morocco and was having them laundered so she could come join you here, if we were never going to send you home."

"Really? You think we'll be here for awhile? You wouldn't mind her coming over?"

"I had to tell her not to bother," David said, his face serious again.

"Oh..." muttered Pat, disappointed.

"You won't be here long enough, Pat."

"Oh?"

"I'm sending you home as soon as this second location is wrapped up. Just for a week or two—so you can check whether Adele packed the right clothes for Bangkok."

"Bangkok?" Pat was suddenly excited, "What's Bangkok got to do with anything? What about this work, and Salif?"

David chuckled.

"Oh hell, they don't need you here anymore. You've got two good crews running things here and at Salif. They'll do alright with this work, plus we'll be moving in some extra men and equipment as soon as Harry gets his mine plans worked out. I've got something much more interesting for you—and it includes a ticket for Adele if you want to take her along."

David explained the basics of the COREX contract with the East Asian Mining Syndicate and told Pat roughly what sort of information the company wanted him to collect on his scouting trip to the Far East.

"You can get all the details from Jack and Norm on the plane back to Vancouver. But that's a couple of days away," David dismissed the subject with a gesture of his hand, "Tell me some more about the Queen of Sheba."

Turning their backs again on the lights of the party, the two men searched the eastern sky for a sign. Pat began his story in a very soft, excited voice: telling David how the Queen went to visit Solomon in his palace at Jerusalem and of the riddle that she posed to the king.

A Short History of Diamond Drilling

From earliest times, mankind has treasured the ability to drill holes in rock. Stone objects with neat, smooth holes through them were regarded as things of value—to be used as a medium of exchange, to be strung along a piece of rawhide and worn as jewelry or attached to wooden handles and used as tools. It is known that certain lower forms of animal life have the natural ability to drill holes in rock, and perhaps man took his cue from one of these creatures. Certain varieties of molluscs drill mechanically by rotating the edge of their shells against rock. Other creatures use chemical secretions to 'dissolve' holes in stone. Or it could be that man first became entranced with the idea of rock drilling by watching the natural processes of erosion at work. Certainly this must be how man came to understand the important role that fluids play in rock drilling.

By whatever means, civilization came to value and systematize the technology for rock drilling. There was always something that man wished to set into or extract from rock. In more recent times, mankind became irritated that large masses of rock stood in the way of completion of time-honored dreams· and schemes. Ever larger rock masses were approached with the idea of drilling holes in them. In 1857, two construction crews approached the base of Mont Cenis in the European Alps with the intention of blasting a hole through that chunk of rock to connect France and Italy by rail. Our diamond drilling story really begins here, with this challenge, but first let us look back in time.

The ancient Chinese were known to have developed hand-drilling techniques to gain access to underground water supplies. The Egyptians took rock cores of short lengths during construction of the pyramids. Using twist drills, probably of a bow-type, they used abrasive powders and hard pebbles as their drilling mediums. Various simple hand-operated drills, augers and churns made their appearance throughout history. Into the 20th Century, hand-drilling in certain types of rock was still accomplished with the pointed

77

steel wedge and hammer. These simple tools were used to drill blast holes for the placement of black powder. With the advent of the Industrial Age and the development of power mechanics, modern drilling technology was born.

The Industrial Age, or Machine Age, marked the transition from mankind's use of tools to the use of machines. Tools are regarded as implements held in the hand and worked by hand to assist in performing certain physical tasks. Machines are devices that augment or replace human effort for the accomplishment of these physical tasks. For 3,000 years, civilization worked with tools derived from six simple configurations: lever, wedge, screw, pulley, inclined plane and the wheel and axle. Beginning roughly around the year 1800 A.D. and continuing into the mid-20th Century, these tools were replaced by machines operating from a variety of power sources. All machines until recent times were derived from combinations of these seven classes: turning machines, milling machines, drilling machines, grinding machines, shapers and planers, power saws and presses. Drills then are one of the basic types of machine, and drills played an extensive role in converting our society from tools to machines.

Thomas Newcomen, a British inventor, offered the world a low pressure steam pump, an 'atmospheric engine', in the year 1712. Forerunner of Watt's steam engine, the device was used to pump ground water from working mines. Ninety years later, another British inventor, Richard Trevithick, developed a sequence of steam-powered machines just as the Industrial Revolution was beginning to sweep the world. Following the lead of Newcomen and Watt, Trevithick unveiled a high pressure steam engine in 1802 as a source of inexpensive power. Within two years he developed a locomotive and a steam-driven carriage that ran on rails. Turning his attention to industrial problems, he pioneered the power thresher and developed a number of machines for mining, including a steam-driven rotary rock-drill in 1813. Roughly fifteen years later, Isaac Merritt Singer in the United States invented a steam churn drill. He went on to make his name a household word with America's first consumer appliance, the Singer sewing machine, in 1851. The piston drill, with a power stroke, made its appearance in 1843. Ten years later, a drill resembling a modern air drill made its appearance in Germany. Many of these early drills were more complicated than reliable and one source reports that it was often necessary to stock 200 units on site to ensure that there were 16 drills in constant operation, all the rest in various stages of repair.

Napoleon III was emperor of France during these times and the Industrial Revolution had a dramatic effect on that state. A railway law was passed in the year 1857 to provide a transport system across the country and one of the first projects initiated with government financial backing was the Mont Cenis Tunnel. Begun that same year and completed in 1871, the tunnel is eight miles in length and remains a major engineering feat of the mid-19th Century. It was the first long-distance tunnel driven from two headings with no intervening shafts. Work progressed from Bardonecchia in Italy and from Modane, France, to complete a through track from Chambéry to Turin, running beneath the Col de Fréjus.

During the fourteen years of its construction, the technology and the hardware of rock drilling underwent a massive evolution at the site. The first few years of construction saw the drilling being done by hand at a rate of 9 inches/day at either end of the tunnel. Compressed-air was introduced as a power source in 1861 by Daniel Colladen of Geneva, but while the pneumatic percussion drills increased the drilling rate five times, the operating costs were beyond the economics of most contractors on the project. It was at Mont Cenis that dynamite first replaced black powder as the explosive for construction use. An improved rock drill was developed by project director Germain Sommeiller, but it was soon overshadowed by the birth of the diamond drill.

The first modern diamond drill was the brainchild of a

Swiss engineer, Jean Rudolphe Leschot, who made his home in Paris and was working on the tunnel. He called his invention a 'perforator' because it was designed to drill blast holes. He built the prototype during 1862-3 with the aid of his mechanic, Pihet. Transmitting roughly one-half horsepower via belt drive to the drill-head, this first diamond drill was set with a ring of black diamonds and could bore 4.6 feet/hour [1.37 m] through granite, producing a hole of 1.38" [3.5 cm] diameter.

It was simple logic that brought diamonds and drilling together. For thousands of years, diamonds had been used as cutting tools but their rarity and expense prohibited their use in any volume. They had been used before for drilling, in a handbar drill illustrated in Diderot's 1751 Encyclopaedia of Sciences, Arts and Trades, and perhaps this device was an inspiration for Leschot. The dramatic increase in the rate of productivity of drilling equipment subsequent to the introduction of power drills virtually assured the replacement of the relatively soft metal bits with diamond-studded bits for hard rock drilling. As the hardest known materials, diamonds were the natural choice for obtaining the highest productivity rate. With the development of a steady supply of inferior grade stones from colonial sources at reasonable prices, diamond drilling rearranged the economics of hard rock drilling to everybody's satisfaction.

Within a year after the introduction of his first 'perforator', Leschot had new models of his steam-driven diamond drills at work at Mont Cenis.

The British also began experimenting with the idea when word circulated about Leschot's invention, and in England as well as France, diamond drilling was at first used almost exclusively for blast hole drilling. Colonel F.E.B. Beaumont began his work with diamond drills as a junior officer in the Royal Engineers. In 1868, he was granted what was perhaps the first patent for a major improvement of the Leschot drill. He was awarded another patent in 1872 with a partner,

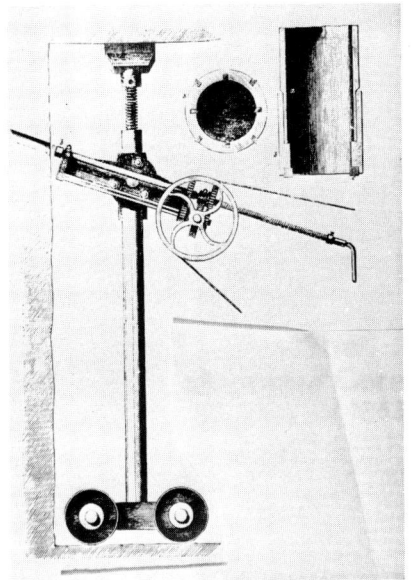

Boyles CORE BOX Feb. 1963

The original diamond drill designed and built by Jean Rudolphe Leschot in 1862 was a very simple device. Called a 'perforator' because it was designed for drilling short blast holes, it consisted of a direct gear driven mechanism to rotate the drill shaft. The shaft with its diamond-studded bit was advanced into the rock by a hand-feed screw. Power was transmitted to the gear drive via a belt system that coupled with a variety of power sources or engines. It could bore 4.6 feet/hour through granite producing a hole of 1.38" diameter.

Col. F.E.B. Beaumont of the Royal Engineers made major improvements on Leschot's device and developed numerous other machines using the diamond drill as his model. In 1875, he was involved in an ambitious project to tunnel under the English Channel to provide England with a permanent connection with the European continent. He designed this tunnelling machine for the work and shipped its twin to France to begin construction of the tunnel at that end. Almost half a mile of tunnel was completed from each direction before the project was abandoned and the machines left to rust in the tunnels.

Science Museum, London, England.

Joy Mfg. and the Newcomen Society

Boyles CORE BOX Feb. 196

Albert Ball began his career with the Sullivan Machinery Company of Claremont, New Hampshire, with his invention of this rail-mounted diamond-tipped quarry channeling machine for use in the marble quarries of New England in the early 1870's. In 1872, a steam-driven diamond drill which may have been another of Col. Beaumont's creations was photographed at work at Silver Islet, Ontario, near present-day Thunder Bay. That same year, diamond drills which may have been produced under Leschot's direction were used in a massive harbor-clearing project designed to rid the waters around Manhattan Island, New York, of various navigational hazards. Shown below are two different types of bits used on the project. The solid bit studded with black diamonds was used to drill blast holes. The hollow bit was a coring bit based on Leschot's design. It was also used for drilling blastholes and for testing rock formations in advance of tunneling crews.

MFG. & BUILDER mag., July 1872

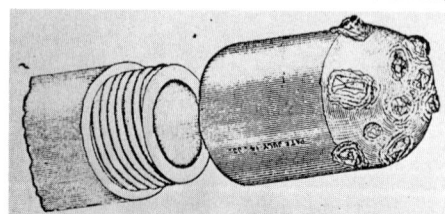
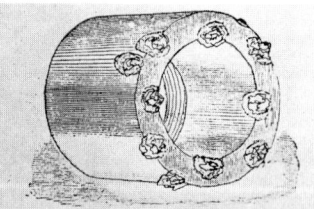

The steam-driven Sullivan diamond drill of the late 1800's was a relatively simple device which was reliable and simple to operate. These drills remained the workhorses of the diamond drilling industry for many decades. The basic concept of diamond drilling has not changed significantly since Leschot invented his first "perforator" in 1862, regardless of the equipment used—its motive power or choice of drilling medium. The drill's engine powers the head which rotates the rods to grind through the rock. The pump supplies water or mud or another fluid to the diamond bit to cool the bit as it drills and to flush the ground rock from the drill hole. The hollow drill rods attach to a core barrel which accepts the rock core in specified lengths. When the core barrel is full, the string of drill rods is hoisted to the surface for recovery of the core. The drill hole is cased in soft or broken ground to insure that the direction of the hole remains "true".

Boyles CORE BOX Feb. 1963

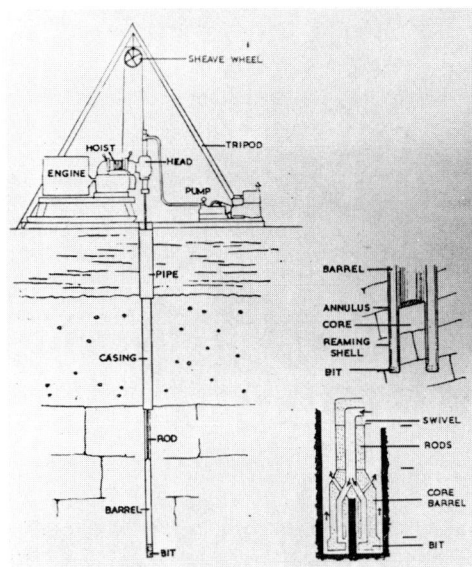

K. McGregor
THE DRILLING OF ROCK, 1967

MUSSENS LIMITED, MONTREAL

SULLIVAN DIAMOND DRILLS.

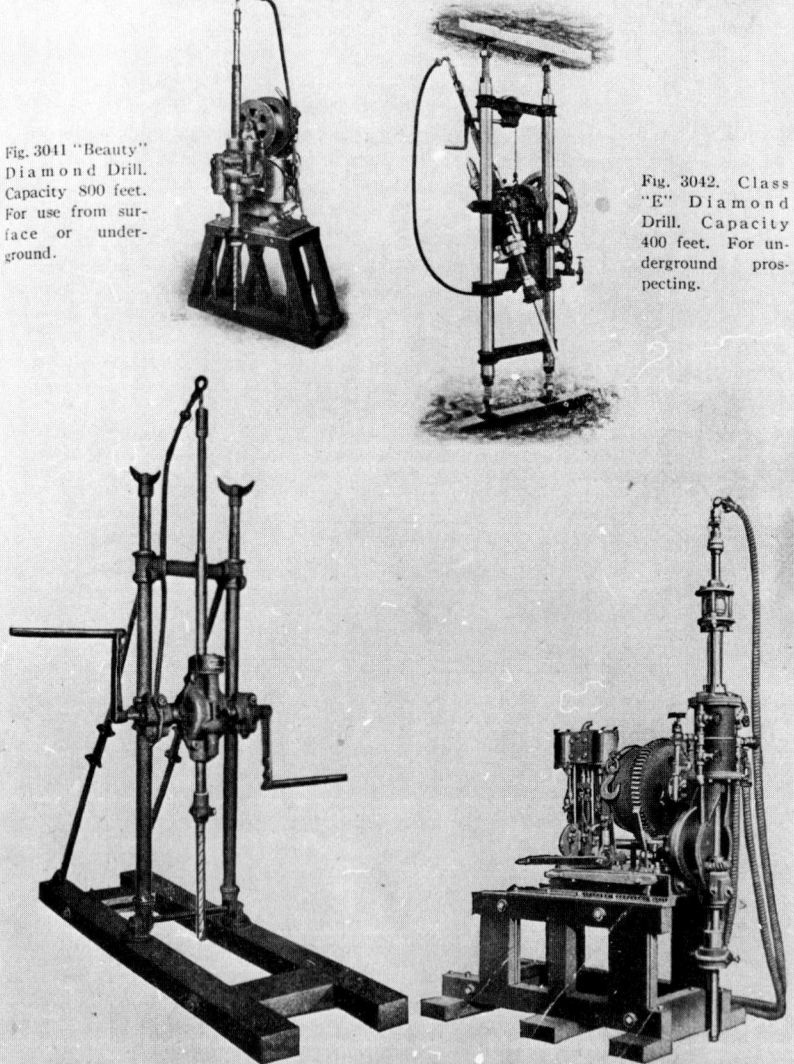

Fig. 3041 "Beauty" Diamond Drill. Capacity 800 feet. For use from surface or underground.

Fig. 3042. Class "E" Diamond Drill. Capacity 400 feet. For underground prospecting.

Fig. 3043. "Bravo" Hand Power Diamond Drill. Capacity 300 feet. The capacity of this drill may be slightly increased and more rapid progress made by the attachment of a horse-power or by operation by belt from a gasoline engine.

Fig. 3044. Class "H-2" Diamond Drill. Capacity, 1,000 feet. This is the standard type of Sullivan Diamond Drill, with hydraulic feed, vertical engines for steam or air, and separate hoisting drum. This style is built in capacities from 1,000 to 6,000 feet.

Jim Cullinane, Sr.

Illustrations of major types of Sullivan drills in the 1913 Mussens 'Blue Book' catalog of railway, mining and contractors supplies.

Joy Mfg. and the Newcomen Society

"BRAVO" ELECTRIC DIAMOND CORE DRILL—EARLY 1900'S

In July of 1913, Mussens of Montreal published their 50th catalog of railway, mining and construction supplies. The 'Blue Book' contained an illustrated selection of Sullivan diamond drills and a photograph of an electric diamond drill working underground in a Mexican gold mine. The 1000-ft. capacity Sullivan electric drill is shown boring a 45-degree upper hole. The Blue Book contains this description of diamond drilling technology:

The diamond drill, during the last thirty years, has proved itself indispensable to the prospector and mine manager. It consists of a line of hollow rods, rotated by an engine and carrying, at its lower end, a bit or crown, set with black diamonds, or carbon, which grind away an annular hole in the rock. The core within this hole is at intervals withdrawn, forming a perfect record of all solid materials penetrated. It is thus possible to determine the exact distance of the body of ore or other material sought, from the point at which the drill is started, the thickness and quality of the ore, and the character of the material penetrated in reaching the ore body. It is, therefore, possible to learn the exact cost of development work, without going to the expense of sinking a test pit or shaft, or running tunnels or drifts. Engineers and contractors find the diamond drill of great service in making test borings for dams, tunnels, bridge foundations and other similar structures, which require an accurate knowledge of the materials to be encountered before work is commenced.

The drill in use is similar to the Sullivan "Bravo" electric drill.

Jim Cullinane, Sr.

U.B.C. Library, Special Collections division.

The diamond drill moved into the wilderness areas of western Canada as the handmaiden of the corporate mining establishment—to provide precise geological information to a professional industry. In earlier times, prospectors like "Scotty" Mitchell, who roamed the Cariboo district during the 1880-90's, explored the remote areas of British Columbia looking for signs of mineral wealth. When they made their discoveries, word spread and other men came to develop mines of one sort or another. Most mineral claims of those times were worked with relatively crude equipment and by simple methods. This placer mine at Atlin, B.C., is fairly typical of the mining activity of the time.

U.B.C. Library, Special Collections division.

U.B.C. Library, Special Collections division.

With the discoveries in the 1890's of massive orebodies such as those at Rossland and Kimberley, the mining scene changed. Deep underground mines became the order of the day. The landscape was dotted with massive structures containing the lifting equipment to run the mine shaft elevators and hoists, and some of them became monuments for their time like the head works of the famous War Eagle Mine at Rossland. Underground, the work progressed by whatever means proved economical. Before certain mining companies could afford to equip their miners with percussion drills and more sophisticated gear, much of the work was done by the hand-drilling method. In this view of the face of the tunnel at the Cliff Mine at Rossland, two miners drill blast holes by the hand drilling method. One miner holds a spiked rod and twists it in the hole while the other bangs it with a sledge hammer.

U.B.C. Library, Special Collections division.

Just as the lumberjacks had their logging festivals and the cowboys their rodeos, the miners of the Kootenays had their 'Sports Days' to socialize and compete against one another for honors and cash prizes. Partners who worked together in the mines would compete together as teams. McDonald and McGilvery were famous in their time for being one of the best hand drilling partnerships and here they are shown competing in a hand drilling contest in Sandon, B.C. in 1904. Sports Day included carnivals, games of chance, displays of food and livestock and horseraces like this July 1st celebration in Nelson, B.C.

U.B.C. Library, Special Collections division.

R.H. Trueman & Co. Photography.

Jim Cullinane, Sr.

Vancouver Public Library photo archives (Dominion Photos collection)

Sports Day evolved along with changes in technology. Percussion drilling contests replaced hand drilling contests and the festivals became more organized. Jim Cullinane's father and his partner were known as the foremost percussion drillers at the Centre Star Mine in Rossland and won numerous prizes competing as a team in the percussion drilling contests of Sports Day. Percussion drilling has always been the primary means of drilling blastholes underground, displaced for only a very short period of time when the economics of the diamond drill became more attractive for this purpose. Here a percussion drilling crew drills blastholes underground at Alice Arm. One man feeds the drill shaft with a hand crank. A piston within the drill head hammers the shaft while it is rotated in the hole by the pneumatic cylinder.

Boyles CORE BOX 1958 B.C. Centennial issue.

Elmore F. 'Bill' Boyles: 1864-1928   Page Boyles: 1866-1917

The brothers Elmore "Bill" Boyles and Page Boyles founded the company which was to become one of the greatest diamond drilling contracting and manufacturing firms in the world. They began in 1895 with one drill and a contract at the Republic Mine in Washington State. In 1926, the company was sold to L. Jessen and F. Lindhe, who developed an international reputation and expanded the manufacturing operations. The Boyles Bros. shop of 1933 produced the portable X-RAY for prospectors, the BBS-1 surface drill and the lightweight BBU-1 "Baby" for underground drilling.

W.J. Moore, Vancouver.

Boyles CORE BOX Feb. 1963

In a period of a few short years in the early 1930's, Jessen developed a variety of drills which were to become the mainstays of the manufacturing division of Boyles Bros. Drilling Company. The first machine he built was composed chiefly of parts from a Model T Ford and a drill head from a steam drill by Longyear. Weighing 1400 lbs., it could drill 1500 ft. It was skid mounted and could be moved through the bush using its own hoist and cable to "donkey" it along. The X-RAY drill weighed just over 100 lbs. and could be packed in to remote sites by a strong man or airlifted in the bush planes of that era. He developed a light-weight underground drill, forerunner of the BBU series, and this skid mounted drill which weighed in at only 550 lbs. The prototype of the world famous BBS-1, it had a drilling capacity of almost 600 ft. and recovered a core of 15/16".

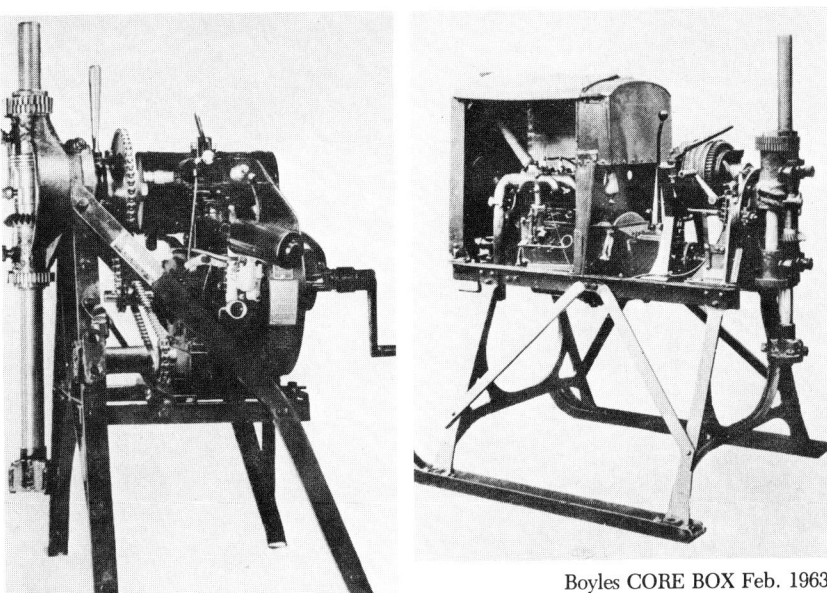

Boyles CORE BOX Feb. 1963

Boyles CORE BOX Feb. 1963

Underground drills went through an evolution at the hands of L. Jessen and were reduced in size and weight by about half in a period of less than ten years. The air-operated Boyles Bros. XRA drill used by a Connors Drilling crew at the old Nickel Plate Mine at Hedley in the late 1930's was the direct descendent of the old BBU-1 drill used at Surf Inlet in 1923 by a crew from Boyles' contracting service. Many fingers were lost to the gear drives of the old drills before gear covers were made an integral part of the casting for the drill head. Technological advances inspired the development of today's more sophisticated machines, such as this drill in operation underground at Downie.

Surf Inlet

Vancouver Public Library photo archives.

Hedley   Jim Cullinane, Sr.   Downie   Arne Rosen.

Jim Cullinane, Sr.

Exploration drilling has always been regarded as the most colorful and romantic aspect of diamond driling. In the early days it was truly a 'frontier' occupation. Horses were just as accustomed to packing drill rigs into the mountains as they were to chasing cows. The horse was the exploration driller's constant friend and companion. In 1925, young Jim Cullinane went north as part of an exploration team to the Hay River district. Each man on the expedition had a horse and toboggan to guide. Pack trains remained an important element of the driller's life until the wide-spread use of aircraft finally replaced the horse. Slow as they might be, the horses could at least keep moving in weather that would ground most aircraft.

Jim Cullinane, Sr.

Jim Cullinane, Sr.

Lead horse of a pack train to Yalakom in the Bridge River district in B.C.'s southern Coast Mountains.

One advantage of being the lead horse of a pack train is that you get the first unobstructed view of the country the drillers venture into. Sometimes spectacular, sometimes forbidding, the view foreshadows the requirements of the job. The trail to Yalakom in the Bridge River district exhibits the remote and rugged qualities of the Coast Mountains which make this region one of the most difficult to travel on the North American continent. Farther north, the country becomes wilder still.

Jim Cullinane, Sr.

Mountain camp at George Copper in the Portland Canal region in the northern Coast Mountains.

In 1928, a crew from Connors Drilling traveled to the Portland Canal region to sample cores on properties owned by George Copper. Portland Canal is a fjord separating B.C. from Alaska with the town of Stewart at its head. The drillers' base camp was established in the mountains at an elevation of 3,000-ft. Drilling platforms were constructed of timbers hauled up the mountain from below and positioned in places where no man should be able to walk upright. The ore from some of the mines in the Portland Canal area had to be packed out of the mountains on horseback.

Jim Cullinane, Sr.

Drill rig built of timbers on mountain at George Copper in the Portland Canal region, 1928.

Jim Cullinane, Sr.

Drill rig on the mountainside at George Copper, 1928.

Drillers on their day-off visiting a neighbor with a jug at the hot springs near Skookumchuck, B.C., 1926.

Jim Cullinane, Sr.

Every so often drillers in remote areas will take a day off to explore and visit with their neighbors. This lucky group of drillers working near the hot springs at Skookumchuck, B.C. in 1926, found some fellows with a jug. One Sunday in 1929, a Connors Drilling crew exploring a deposit at Babine Lake came down to Topley Landing to the place where Daniel Leon was setting up his little community. The drillers all took turns whipsawing lumber in order to tell their heirs they had done it once before the technique passed into antiquity. Some of Leon's buildings remain in use there today.

Jim Cullinane, Sr.

Drillers trying their hand at whipsawing lumber at Babine Lake, 1929, before the technique passed into antiquity.

Dog team moving drill across frozen lake near Goldfields, Saskatchewan in 1938.

Jim Cullinane, Sr.

Dugout canoe in use on Pinchi Lake during the 1940's.

Jim Cullinane, Sr.

Drill crews in remote areas must depend upon whatever means of native transportation they can find to move their equipment and travel around the country. In the 1940's rush to develop the mercury deposits at Pinchi Lake in northern B.C., a drilling crew found itself using dugout canoes made by Indians in the area as the main source of transportation. At Goldfields and Dinty Lake in northern Saskatchewan, dog teams proved to be the most reliable transportation available to move men and drills.

Jim Cullinane, Sr.   Dinty Lake, Saskatchewan

Jim Cullinane, Sr.

The skiplane soon replaced the dog team as the major source of transportation into the remote areas of the north.

Then there came the day in the 1930's when the dog teams at Dinty Lake in northern Saskatchewan met the first airplanes equipped with skis. The airplane revolutionized transportation in the north within a few short years. Bush planes equipped with skis displaced the dog teams, and airplanes with floats displaced the boats. Traveling by air, drill crews were able to invade the most remote areas of the country like the area of McDame Creek on the British Columbia/Alaska border.

A drilling crew traveled by ship from Vancouver to Stewart, B.C., then by plane to the glaciers of the McDame Creek area of the northern Coast Mountains. Even in high summer most of the area remained under snow. The drillers were fortunate when they could place their rigs directly on rock. More often than not, they were forced to dig through the snow and then through an overburden of sand and gravel in order to position their rigs to drill a true hole. This early BBS-1 had a straight 4-cylinder Wisconsin engine.

Jim Cullinane, Sr.

McDame Creek, Alaska.

Jim Cullinane, Sr.

A BBS-1 drill positioned under the snow to drill steady into rock.

There were always places where the airplanes could not go—where horses and dog teams and dugout canoes were of little use. That is when the diamond driller himself became the beast of burden. From the Rocky Mountains to the Pacific coast, the waterways of British Columbia carve spectacular canyons from the mountain slopes. The rock structures of these canyons must be surveyed and tested prior to the construction of dams and hydro-power developments. It is the diamond driller who gets to climb down into the canyon and traverse the river in search of this geological and engineering data.

Jim Cullinane, Sr.

Moving a Knight & Stone 'Sidewinder' in Phillips Canyon 1930.

Jim Cullinane, Sr.

Jim Cullinane jockeying drill across the Salmo River.

Jim Cullinane, Sr.

Drill rig on raft at bottom of Phillips Canyon near Elko, B.C.

Jim Cullinane, Sr.

Drill raft anchored in Elk River, foundation testing for a dam site.

In some cases, the drillers must work on the water itself, to extract core samples from the riverbed. One such job undertaken by a Connors Drilling crew in 1930 involved drilling from a raft in Phillips Canyon on the Elk River near Fernie in the Rocky Mountains. A Knight & Stone 'Sidewinder' drill was packed down into the canyon, set up on a raft and then moved out on the river to be anchored with overhead cables. The work was cold, dangerous and exhausting.

John W. Booth.

During the Second World War, the Boyles Bros. manufacturing division held the contract to supply diamond drills to the various Allied army engineering corps. This wartime view of the Boyles' Vancouver shop shows a number of drills being readied for shipment. Technological advances made during the war inspired the development of more sophisticated machines, which helped Boyles to assume a position of supremacy in the manufacture of diamond drills in Canada during the 1950's and 60's. The advertisement indicates the extensive line of materials which Boyles supplied to the industry.

Boyles CORE BOX
B.C. Centennial Issue 1858-1958.

Boyles Bros. advertising material.

During the first half of the present century, Ripple Rock in Seymour Narrows was considered to be the most treacherous navigational hazard of the inland waterways of North America. Even powerful vessels such as Union Steamship's T.S.S. Cardena, which ran a regular schedule along the B.C. coast, were made to wait for slack tide to maneuver around the Rock's whirlpools and eddies. The Boyles Bros. Drilling Company of Vancouver was a major contractor on the project to tunnel beneath Ripple Rock to pack the twin peaks with explosives for their demolition.

U.B.C. Library, Special Collections.

The demolition of Ripple Rock took place April 5, 1958. One of the photographers to cover the event was R.E. Olsen of Vancouver, who used a motorized sequence camera to capture these shots of the largest non-atomic blast ever produced by mankind.

E. Olsen

The distance across the channel at this point is roughly half a mile. Vancouver Island is to the left in this sequence of photographs, and Maud Island to the right in front with a point of Quadra Island behind. The cloud of debris from the explosion rose well over half a mile into the air.

COMINCO.

Aerial photograph of the west coast of Greenland with Marmorilik Point at center right.

The development of the Black Angel Mine on the west coast of Greenland was an extraordinary project. The lead/zinc deposit was discovered on the face of 3,000-ft. high Marmorilik Point [right center in the photo] and the base camp for the mine was constructed at the foot of the talus slope on the other side of the fjord. The view up the tram line from the base camp reveals the twin portals of the rock dump with the portals of the mine's main shaft above them. The view is breathtaking from the portal of the mine at 2000-ft. elevation, year-round.

Canadian Bechtel Ltd.

View of the mine portals from base camp.

View from the portal of the main shaft of the Black Angel Mine, Marmorilik Point, Greenland.

COMINCO

COMINCO.

Drill rig on a Bombardier half-track vehicle on muskeg bogs at Pine Point on the south shore of Great Slave Lake, N.W.T.

A drill rig mounted on a Bombardier half-track vehicle makes exploration drilling more productive in the muskeg bogs of the Northwest Territories at Pine Point on the southern shore of the Great Slave Lake. The helicopter helped revolutionize exploration drilling by providing an economical means of transporting men and equipment to remote drill sites. A project supervisor, using the helicopter as his main source of transportation, can provide geologists with immediate access to his drilling crews which might be scattered over a wide area of wilderness country.

Connors Drilling Ltd.

Helicopter servicing a drill rig on a mountaintop in northern British Columbia.

Connors Drilling Ltd.

Connors rig in woods.

In the end, this is what exploration diamond drilling is all about: the core samples taken from the earth are placed in boxes and delivered to the geologists for examination. The cores are arranged in their boxes in linear fashion with wooden markers placed at the ends of each row to measure footage. The geologist pieces the core fragments together in the lab and, examining the core inch by inch, constructs a picture of the ground from which the core was taken and makes predictions about the immediate surrounding area. Pieces of ore found in the sample can be removed for assay to determine the economic value of the mineral deposit.

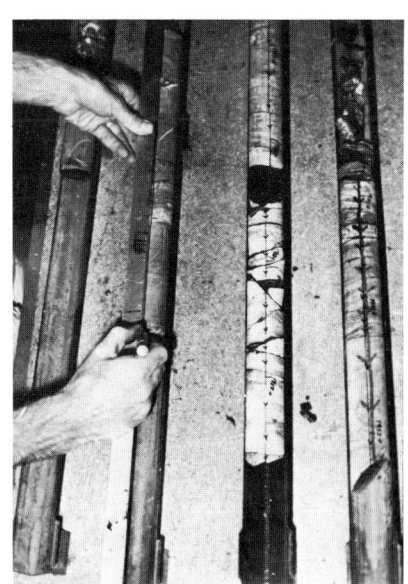

Connors Drilling Ltd.

Geologist marking orientation of core sample.

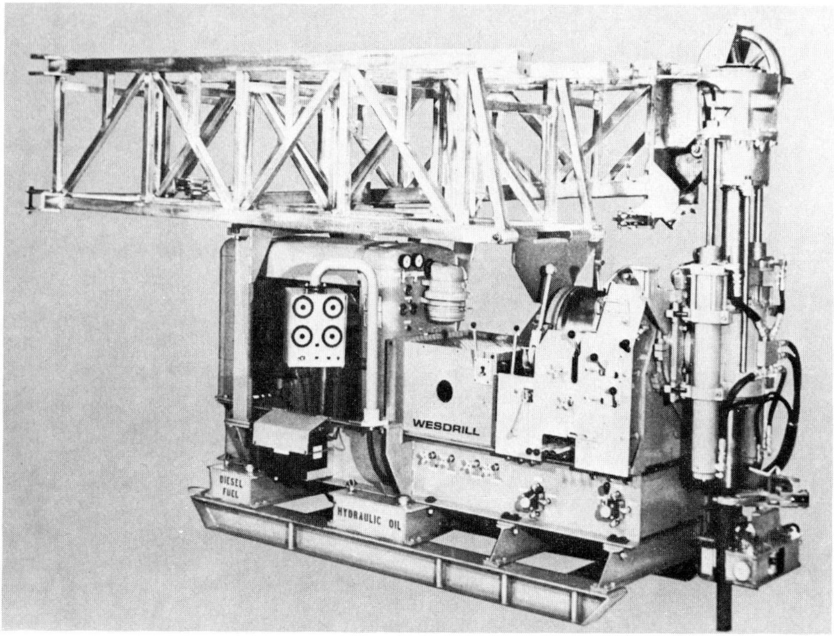

Wesdrill Equipment Ltd.

One of the most recent advances in diamond drilling technology was the introduction of the Wesdrill Model 60 Surface Diamond Drill. The machine can be powered by gasoline or diesel engines or an electric motor, and is fitted with a folding drill tower and the Wesdrill Wire Line Hoist system. The instrumentation of the Wesdrill is simple, uncluttered and easy to read and program.

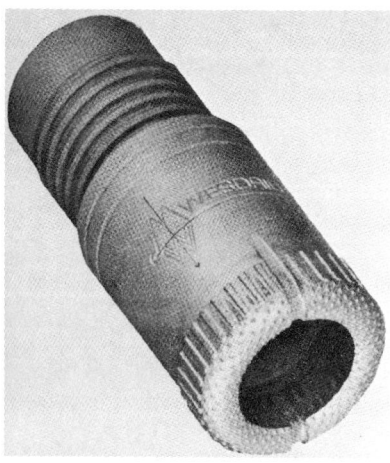

Wesdrill Equipment Ltd.   Wesdrill Equipment Ltd.

ILLUSTRATED ARE SEVERAL STEPS IN MAKING PERMASET DIAMOND BITS AND REAMER SHELLS.

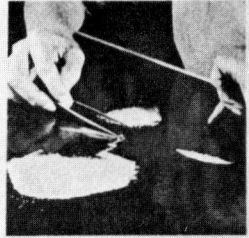

1. SORTING DIAMONDS The Diamond Boart that put the "bite" in today's Permaset drill bits are carefully hand-sorted by experts, who classify the stones according to size, shape and quality.

2. THE PATTERN The performance and durability of Permaset diamond drill bits begins with exclusive diamond-setting patterns — one of which is shown being marked on the graphite mold in this photograph.

3. THE SETTING A quarter-century ago, diamond drill bits were handset with six to twelve large, costly black diamonds. Today's Permaset bits use many smaller, less expensive diamond boarts, using the modern method illustrated above to ensure uniform settings.

4. POWDER AND SPATULA After tungsten powder is packed on top of the diamonds, the bit blank and bonding metal are added.

5. HOT BIT Hot bit is the product of the high temperature furnace. In less time than it takes to bake a cake, the bonding metal infiltrates the powder which makes a chemical union with the diamonds to form a dense, hard crown firmly attached to the steel shank.

6. Here all surfaces of the shank are finish-machined to size prior to the final threading.

7. REAMER SHELLS Reamer shells, casing bits, shoes, etc., are finish-machined. Here the illustration shows a reamer shell being threaded in a lathe.

8. FINISHING TOUCHES Each bit and shell is bathed in cadmium plating to brighten the finish and prevent rust.

Boyles Bros. 1965 annual report.

Diamond bit manufacturing has come a long way from the crude handset bits used to drill blast holes for the clearing of New York harbor in 1872. The photo story taken from the 1965 annual report of the Boyles firm illustrates the process for casting modern diamond bits. The Wesdrill bits are typical of the clean and efficient tools used by today's drillers. J.K. Smit, Wesdrill and the Longyear company remain the largest suppliers of diamond bits in Canada.

No one can be sure what changes will occur in the diamond drilling industry in future years. It is likely that automation will make core sampling more precise a means for obtaining geological information than in the past and it is possible that miniaturization will offer contractors machines that are even more portable than those currently available. Diamond drills with sophisticated instrumentation like the Wesdrill, shown here drilling in the mountains of Colorado, will play a more important role in the near future. Whatever changes may be in store, tomorrow's drillers will owe a debt of gratitude to those who wrote the history to date of the diamond drilling industry in western Canada.

Wesdrill Equipment Ltd.

Jim Cullinane, Sr.

Appleby, for the design of a drill which was used in 1875 to drill a hole 2,288 feet [687 m] deep. Beaumont designed yet another spectacular drilling machine, a tunnel borer, which was used in 1881-2 during the first serious attempt to tunnel the English Channel from Dover to Calais. Almost 2,000 yards [829 m] were completed at both ends of the tunnel before the project and its exotic equipment were abandoned.

During the 1870's and 80's, diamond drilling became a routine accepted method for blast hole drilling in hard rock and began to acquire prominence in the mining world as the primary technique for core drilling. Drills appeared on the scene with speeds in excess of 360 RPM at 5 horsepower, and a core hole drilled in Germany in 1886 reached 5,734 feet [1748 m] deep.

Across the Atlantic, an independent drilling technology was developing in the United States. The world's first patented rock drill was introduced in 1849 by J.J. Couch of Philadelphia. It was a percussive-type drill in which the bit was mounted on a shaft and thrown like a spear through a hollow piston at the rock, then gripped and returned through the piston. A pneumatic-feed hammer-type drill was introduced later in the century by a Denver machinist, C.H. Shaw. It was through this lineage of Singer, Couch, Shaw et. al., that rock drilling progressed in the United States.

The U.S. had its tunnel, too, in Western Massachusetts. The Hoosac Tunnel was driven through the southerly range of the Green Mountains just to the east of North Adams, Massachusetts. It was the first rock tunnel in the U.S. and it was to provide Boston with access to the West. A 4-mile long railway tunnel, it was begun in 1851 by Col. Herman Haupt and completed in the mid-1870's by the firm of W. & J. Shanley of Montreal, Quebec. Construction was constantly interrupted by financial and political difficulties, but here again the resident technology evolved in dramatic fashion. The resident engineer, Thomas Doane, pioneered the use of nitroglycerin for blasting and the use of electricity to fire this explosive. The Hoosac project also saw the development of compressed-air drilling equipment that launched the American pneumatic tool industry. No mention is ever made that diamond drills were to be found at Hoosac.

However, there is evidence that Leschot built a number of his diamond drills for export to the United States for use on the East Coast. It is probable that one of the Leschot machines served as a model for the first American machines.

In 1867, M.C. Bullock patented a steam-driven drill and used his 250 RPM machine as a core sampling device. He drilled a 750-foot hole that year at Pottsville, Pennsylvania, to map the area's coal deposits.

There is a report in the July 1872 issue of *The Manufacturer And Builder* magazine published in New York, which reveals a great deal about the fledgling diamond drilling industry just ten years after its birth. A subject of major concern to New Yorkers at the time was the rearrangement of tidal patterns in the Hell Gate section of the East River. There were fears that the mouth of the East River was becoming silted over in the manner of the German Rhine in Holland and it was demonstrated that the sand bar of Coney Island, along the south shore of Long Island, was evidence of this trend. There was pressure as well from the shipping fraternity to open a safe route for navigation direct to Manhattan from the protected waters of Long Island Sound. This required the removal of numerous rock masses from the Hell Gate rapids and the removal of certain rocky points along the shore such as Hallet's Point below Fort Stevens in Astoria in Queens County. In 1857, the process of removing the major rock masses from the channel began under the direction the French engineer, Maillefert. This was accomplished by surface blasting. The overall plan to clear Hell Gate was developed by Major General John Newton of the U.S. Engineering Corps and relied upon the use of three new processes: the use of diamond drilling which was sweeping the construction industry as a means to deal with large bodies of hard rock, and the use of nitroglycerin and electric detonation which had just been developed at the Hoosac Tunnel.

A coffer dam was built on the shore below Fort Stevens and ten tunnels driven into Hallet's Point below water level by teams of miners from Cornwall, England, consulting with the Army Engineers. The superintendent of the works was the famed mining engineer, Reitenheimer.

Diamond drills were used to drill blast holes for a tunnel about six feet in height sloping down under water. These were followed by other drill teams who helped enlarge the tunnel to twelve foot dimensions. Each of the ten tunnels was more than 300-feet in length. When the excavations were completed, blast holes were driven into the pillars separating the tunnels and galleries and filled with nitroglycerin. An enormous charge of explosives was placed at the far end of each tunnel and all were detonated by a single electric charge on July 4th, 1873. The rock mass fragmented and dropped an average of twelve feet into the river bed and was then removed by simple dredging of the broken and crumbled rock.

The diamond drills used on the job were produced by the American Diamond Drill Company and used both solid bits set with diamonds and hollowcore bits. The magazine reporter describing the diamond drill reports a savings of more than 50% in the use of these machines to do the job formerly done with pneumatic tools:

"They are extremely simple, both in construction and operation; seldom need repairs.... They are not subject to the constant and destructive shocks of concussion against rock, which disable the best percussion machines so often, and render them so expensive to keep in repair...Every part is equally simple and durable. They produce holes of uniform diameter—a feature of great importance in blasting, as the

force of the explosive material is thereby fully utilized...They are not deflected from a right line by seams and crevices, nor impeded in their progress by the hardest rock...They are not only adapted to shafting, tunnelling, well-boring, submarine blasting and all kinds of rock-excavating...but in their application to 'prospecting', they accomplish most important results...It is in the opening and working of mines, the grading of roads, etc., the sinking of shafts, and the driving of tunnels, that the great value of this drill as a labor-saving machine is most apparent."

Another major landmark in diamond drill technology was the 1878 invention of the Sullivan drill, following Leschot's first 'perforator' by a mere fifteen years. The Sullivan Machinery Company was founded in 1851 as a machine shop in Claremont, New Hampshire. Claremont was the county seat for Sullivan County which took its name from the early New Hampshire patriot, Major General John Sullivan. The company's earliest offerings were new designs of old concepts, to make certain machines more powerful or more compact or with other desirable features. The earliest products were engine lathes, iron planers, paper machines, circular saw mills and the Tuttle Water Wheel. Renamed the Tyler Turbine Water Wheel, this Sullivan item won first prize in its exhibition class at the Crystal Palace fair in New York. More than 3,000 Tyler Water Wheels were manufactured by Sullivan and they spread throughout the world, in many places providing the first power sources other than human or animal exertion. The foundations for the company were well-laid when Albert Ball joined Sullivan in 1868 as superintendent and mechanical engineer. An irrepressible inventor, Albert Ball was to make repeated impacts upon the history of the contemporary world. He designed an engine lathe in the late 1850's for turning sewing machine needles with great accuracy. In 1863, he perfected the repeating rifle, forerunner of the Winchesters. The U.S. government ordered 2,000 of the guns during the final stages of the Civil War, but the war was over before the order was finished. A German observer saw the guns in action and made a deal with the government to buy the entire lot. They were shipped to Prussia where they were copied exactly and manufactured on a massive scale. Using Albert Ball's repeating rifle as their principal infantry weapon, the Prussians annexed nearly a dozen neighboring states, wresting Holstein, Hanover and Schleswig *et. al.* from foreign control. The staple weapon until century's end for the Prussian infantryman, the rifles were instrumental in the Prussian successes during the Franco-Prussian War.

Albert Ball's first invention for Sullivan was a diamond bit channelling machine for quarrying New England's great marble deposits. Moving smoothly along rails, the machine revolutionized the quarry industry. In 1878, with the price of black diamonds becoming prohibitively expensive for use in great volume in machines like the channellers, the Sullivan Company was looking for new products to utilize its expertise in diamond setting. Albert Ball invented a diamond drill with a hydraulic feed independent of the rotation of the shaft to combat the chronic problems of most other machines which used the screw-feed principle. The first Sullivans were sent to the Mesabi region of Minnesota.

In 1881, two drills were sent to a coal mine in Iowa and the mine manager was so impressed with their work that he formed perhaps the first diamond drill contracting company. The company became the Diamond Prospecting Company of Chicago in 1884 with Frederick K. Copeland as president. Using only the Sullivan Company's products, Diamond Prospecting quickly became a nation-wide contracting service and a very powerful influence within the mining industry. Copeland merged his company with the Sullivan Machinery Company in 1894. He remained president of the Sullivan firm until his death in 1928, running a world-wide network of contracting services as an adjunct to the company's manufacturing division.

The test which ensured the Sullivan drill its premier position in the drilling world came immediately after its invention when the first drills arrived on the Mesabi 'Iron Range' of northern Minnesota. Iron mining began in the Lake Superior region in 1846 and by 1873 the annual ore output exceeded one million tons, but in view of later developments it was still in its infancy. The massive Mesabi deposits were discovered at Mountain Iron in the extreme northeastern area of Minnesota in 1887 and claimed by Leonides Merritt and his brothers. Located just to the southwest of the Lakehead (the present Thunder Bay, Ontario), the Mesabi region was to become the major iron ore producer for the U.S. from 1892 until the very high-grade hematite ran out in the 1950's. Lower grade taconite ore is still being exploited there today with more sophisticated extraction methods.

The Merritt brothers organized their company in 1890 to mine these deposits which were found to be laid in horizontal rather than vertical seams. The position and boundaries of these deposits in the granite shield were pinpointed using the new Sullivan drills for core sampling on a massive scale. Knowledge of the horizontal structuring of Mesabi allowed the Merritts to undertake mining by the open-pit method rather than the more costly and dangerous shaft-sinking method. The Mesabi proved so valuable that the mining companies which grew up there around the Merritt claims were all absorbed into John D. Rockefeller's resources empire in 1893. The ore from the 'Iron Range' was shiped from Duluth by ship across the Great Lakes to the steel-making centers of America. Production from the region peaked only twenty-odd years ago, in 1953, with an output of 100-million tons net of iron.

The Mesabi diamond drilling expertise was to play a very important role in the development of mining in Western Canada and 'Sullivan' was to become a household word in the Kootenays for a number of reasons. And while that was taking place, the Sullivan drills were winning fans all over the world, from the gold mines of Mexico to the Panama Canal Zone where they played an important role in the canal's construction.

Perhaps the most dramatic story of their success as the workhorse of the industry comes from the Transvaal, in the south of Africa.

In 1867, a young African boy found a pretty stone on the bank of the Orange River in the Orange Free State. It was a diamond and it caused an incredible 'diamond rush' to South Africa. A large group of mines developed wherever a volcanic 'pipe' was discovered and the life of this quiet countryside was thrown into chaos. In 1888, Cecil John Rhodes brought the various mining properties together in one association to present a common front to the diamond-buying world. The association, DeBeers Consolidated Ltd., is still based in Kimberley in the Cape Province and all South African diamonds are concentrated at the Consolidated buildings and offered for sale at Kimberley House. At the same time that Rhodes was laying the foundations for the fabulous South African diamond industry, events were taking place in the neighboring South African Republic (Transvaal) which would reveal that region's position as one of the richest provinces on the face of the earth.

The Transvaal is a locale where an entire rock series of high mineral content was shifted out of normal position by hypogeic forces. In particular, one bed or 'reef' of conglomerate rock became famous throughout the civilized world. Known as the Witwatersrand (*white water ridge*) or more simply as the Rand, this faulted reef of auriferous (gold-bearing) rock runs almost 62 miles [100 km] in an east-west direction and it is 23 miles [37 km] wide. It includes a gold-bearing vein which has produced nearly 1/3 of the world's gold supply in the past 90 years. Today, 55 mines produce $1-billion in gold annually, 70% of the gold produced by the "free" world.

119

Johannesburg, which is situated almost dead center on the Rand, became the center of this mining district and subsequently South Africa's largest city. Gold was discovered at a surface outcropping in 1884 and the main reef was plotted in 1889 at a depth of about 500 feet by diamond drilling. John Hays Hammond went to the Transvaal to supervise drilling on the behalf of the Sullivan Machinery Company's contracting service. He convinced the major gold mining syndicate which was also led by Cecil John Rhodes that they were sitting on top of a major gold vein. The Rhodes interests, acting on Hammond's advice, sold their claims near the outcropping and bought most of the acreage along the dip of the vein. Hammond brought in Sullivan-supplied contractors from all over the world and massed 250 diamond drill rigs to plot the boundaries of the vein. Their work led to the opening of 125 of the world's richest gold mines.

George Alfred Denny worked on the Rand during these early years, until the area became something of a no-man's land during the Boer War of 1899-1902. To some degree the war was a result of the tensions caused by the ebbs and flows of immigration to these gold and diamond fields. Denny was on the Rand working as one of Hammond's contractors to establish the position of the reef. Upon his return to England, he wrote of his experiences in the first practical guide for diamond drillers: DIAMOND DRILLING FOR GOLD AND OTHER MINERALS, published in 1900 by Crosby Lockwood and Son of London.

Denny's book contains advice on buying diamonds, setting drill 'crowns', estimating work rates in various types of rock and it catalogs every type of Sullivan drill and other makes, notably Bullock drills, working on the Rand and gives their specifications. His descriptions of life on the Rand and of the economics of the fledgling contract drilling industry give us a good picture of the state of diamond drilling in its fourth decade, at the turn of this century. He mentions that while most of the drills on the Rand were Sullivans, there were still many hand drills floating around and that while they were "less than useless" in quartz, they were good for coal exploration. Indeed, Africa's largest coal deposits were later discovered just east of the Rand.

The Sullivan drills on the Rand were powered by steam or compressed air, but Denny makes special mention of an electric Sullivan drill that he had found in use at Aspen, Colorado: with a minimum 3-hp motor, pump, hoisting frame with wire rope drum and a swivel base, it was mounted on a skid for portability. His information from Colorado indicates that this drill had a rate of 45 feet/day of two 10-hour shifts, averaged over five months of work. Exclusive of power, the unit cost 45 cents/foot of drilling, with the cost of wages for two drillers comprising almost the entire cost of operation. An equipment catalog from Musson's of Montreal distributed in 1913 contains a photograph of this electric Sullivan unit working underground in a Mexican goldmine.

Denny's advice on establishing an independent contracting business includes this information on costs:

EQUIPMENT: [*the average rig set-up*]
Sullivan "K" drill	£ 1,800
50-foot derrick	£ 175
Water tanks	£ 50
Core boxes	£ 100
Lean-to [*over drill*]	£ 75
Living quarters, etc.	£ 140
[*circa 1900*]	£ 2,340

LABOR: [*10-hour shifts, 12 shifts/week*]
    Superintendent [*sets bits, responsible
      for core and costs*]   £ 40/month
2 Drill Runners [*run the rig*]   £ 26/month
2 Stokers [*run the boiler, help break rods*]   Natives:
2 Derrick hands [*help break rods*]   65-70 *shillings
                                          per month, plus food.*

OPERATING COSTS:
Fuel: approx. ₤1/ton
Water: assumed to be sufficient on hand for pump delivery.
Carbons: approx. ₤7.10 shillings/carat

Denny advises that in 1896 there was a run on the black diamond supply in the Transvaal that pushed the prices up to ₤14/carat. This was at the height of drilling activity on the reef, and he goes on to warn of unscrupulous activity on that front. He cautions that it is always necessary to test the specific gravity of the stones you buy because some fraudulent dealers sometimes coated ordinary carborundum with black lead and sold them as Brazilian diamonds. Carborundum has a specific gravity of 4.0 as opposed to 3.5 for diamonds. An average of 8 diamonds were mounted on each bit with a total weight from 1 to 3½ carats. A hole was bored in the metal of the bit, lined with copper foil and the diamond laid in place. The soft iron of the bit was hammered into place around the stone. (Black diamonds are round carbon stones.) The metal bits were good for an average of 3 settings of diamonds, then the stones were transferred to a new metal crown. The bits, reset three times, were good for perhaps 100-feet of drilling in rocks with a hardness of 7 Mohs.

Denny recommends that 'nuggets' with well-defined angular projections are the best stones to use in the bits. 'Flat' stones or chips had a tendency to 'fly', and the 'Boort' stones from the neighboring South African fields were so crystalline in nature that they would fly and chip in these bits. They were used only when the Brazilian black diamonds became prohibitively expensive.

A comparison of contract pricing by Denny reveals that at this time, the diamond drill had a three-to-one advantage over metal rock drills for the same work. In extremely hard quartz on a mining property where a mill was already in operation and certain basic services provided the drillers, the metal rock drills cost 55 shillings/foot for 500-feet at a rate of 20-25 feet/week = 25 weeks for a total of £1,375. A diamond drill rig costs 20 shillings/foot for 500-feet at the rate of 300 feet/month = 7 weeks work for a total of £ 500. That produced a three-to-one advantage in total cost and the same advantage for information productivity over a set period of time. Clearly then, the day of the diamond drill had arrived for exploration work.

Diamond Drilling in Western Canada

I

The first diamond drill reported at work in Canada was used to explore a coal deposit at Harper's bore-hole in Springhill, Cumberland County, Nova Scotia in 1871. The machine quickly moved to the west, and photographs exist of a steam-powered diamond drill in operation at the Silver Islet Mine near Port Arthur, Ontario, the next year—1872. A Beaumont/Appleby core drill was imported from England in 1875 by the Vancouver Coal Company to drill on their site near Nanaimo, British Columbia. Use of the diamond drill grew steadily, especially in Eastern Canada, but it was twenty years before the diamond drill was to find itself in the right place at the right time for the birth of the modern-day Western Canadian mining industry.

During the mid-1800's, when the Americans began pressing their land claims toward the Pacific region into the Oregon territory, the Hudson's Bay Company acquired a comprehensive lease on the colonial territory of Vancouver Island to formalize the HBC presence in the area. The British Parliament was reluctant to extend that lease to the mainland colony of British Columbia before the HBC produced results in their efforts to colonize the Island. Life on Canada's Pacific coast revolved around fur trapping and trading, fishing and logging. Some of the Scots settlers worked coal deposits near Nanaimo and at the north end of Vancouver Island which the HBC established in 1852 to supply heating fuel for the settlements of the Island and fuel for the paddle-wheel steamers that began making regular stops to these sister colonies. Except for this limited amount of coal mining, there was no organized mining in British Columbia until after 1860.

There was, however, a great deal of prospecting going on as explorers and frontiersmen slowly mapped the expanse of the territory. In the 1820's an outcrop of lead was found on the east shore of Kootenay Lake by HBC trappers and their Indian guides who used it to replenish their supply of rifle shot. This same outcrop was rediscovered in 1868 by an American prospector, Henry Doane, who worked it for awhile until the economics of its location made further development work unfeasible. In 1882, Robert Evan Sproule came to the Kootenays fresh from his discovery of the Granby ore deposits in the Boundary district to rediscover the lead deposit once again. Thomas Hammil, working as a mining scout for American interests, jumped Sproule's claim in 1883. Sproule won his lawsuit but could not pay court costs. The court ordered his property sold and the only people buying were backers of Hammil. Enraged at the injustice of the situation, Sproule took the law into his own hands and shot Hammil in 1885. He was tried and hung for murder in a situation which had sensational repercussions throughout the mining community of the East Kootenays. Other developers moved into the property and met with modest success, but it was not until the century's end that the property originally marked for rifle shot eighty years before was discovered to be the tip of the Bluebell iceberg at fabulous Riondel.

As early as 1833, gold was found in the Interior of B.C. David Douglas, a botanist exploring the region surrounding Okanagan Lake, found traces of gold in a stream entering the lake. He collected enough metal to make a seal with his signet. The find was noted in his journals and dismissed. Twenty-five years later, traces were again noted in the Okanagan stream, but nothing was done to explore further upstream.

The year 1851 marks the beginning of some restless movement on the part of the HBC to develop mining on the coast. Indians from Moresby Island in the Queen Charlottes had been using small gold nuggets which they found on the beaches for trade with the Hudson's Bay Company for a number of years. An expedition was mounted to Mitchell Harbour in 1852 to search for the source of this gold. The Whites on the expedition reported some friction developing with the Indians who resented the intrusion. The friction soon developed into animosity and it was not possible to continue the expedition at the time, so the ship and crew and expedition returned to Victoria.

Eighteen fifty-two also marks the date when the HBC made a very positive move to open the B.C. Interior for mining. Indians of the Thompson River district trading at Kamloops occasionally used gold dust as a barter item. The factor of the Kamloops KBC post encouraged the Indians to prospect by providing them with the elementary tools and with instructions for their use. This encouragement was to pay huge dividends within three years when Indians from the Thompson River region reported a gold find in the area. The circumstances of the report were quite unusual. Reticent in their dealings with Whites, B.C. Indians were quite vocal in their socializing with other Indian groups. Some of the Thompson River band travelled to Walla Walla, Washington, in 1855 to visit a woman of their band who had married a French-Canadian and established social contacts with the Indians in the Columbia River region in Washington. Shown gold taken from the Columbia at Colville, the Indians casually mentioned that it looked the same as the stuff they took from the rivers at home and offered to show their finds to these friends if they came to visit. Four halfbreeds of Canadian lineage visited the area and discovered a major source of gold on the Nicoamen River just north of the junction of the Thompson and Fraser Rivers. Word of their discovery in the summer of 1855 took nearly one and a half years to filter to the civilized world and San Francisco in particular where thousands of late-comers to the California gold rush sat waiting for something else to break. Tens of thousands of men from every corner of the globe were soon in motion heading for the Fraser River of British Columbia. At the time of the first discoveries on the Fraser, Vancouver Island had a resident population of perhaps 450 people with 300 settled in Victoria. There were perhaps 300 whites living on the mainland, most of whom worked with the HBC, and a total Indian population for the province of roughly 15,000.

With thousands of men spilling into the region lusting for gold, friction soon developed between the newcomers and the native peoples. This drew a swift response from James Douglas, Chief Factor of the Hudson's Bay Company in the West. Leaving his base at Victoria with the troops from the Victoria stockade, Douglas went to the mainland to restore order. Sir Edward Lytton, Britain's Colonial Secretary at the time, made a very shrewd move when he was informed of the troubles. He cancelled the HBC leases outright and appointed Douglas the Governor of the mainland colony and Governor of Vancouver Island.

Later that year of 1858, gold was discovered in the Cariboo country near Quesnel. This second gold rush brought so many people to the territories that for a time, the city of Barkerville in the Cariboo was the most populous city north of San Francisco and west of Chicago. Governor Douglas, seizing his opportunity, began construction of major roads to connect the various population centers of the two territories. A road from Harrison Lake to Lillooet was completed in 1859 and the 'Great North Road' from Yale to Quesnel was finished in 1863. The Dewdney Trail was pushed deep into the Boundary and Kootenay districts to encourage Canadian settlement to control the sudden influx of American drifters into that region.

Most of the men involved in these ventures were there only for the surface gold that was obtained with simple tools like the gold pan and the sluice. Few of them were prepared to undertake or finance a major mining project to extract gold from underground veins. When the surface gold began to run out, many prospectors left the area and the colonies suffered their first mining depression. The resident population of both the Vancouver Island colony and the British Columbia mainland was roughly 10,000 people in 1860, evenly split between the two territories. They were combined in order to survive the economic strains of this period and the new province of British Columbia got Victoria as its capital.

In a period of twenty years, from 1858 until almost 1880, an average of 1,500 miners and prospectors retrieved $35-million in gold from an area of only 50 square miles in the Cariboo district. Those who followed the first frantic rush of 1858 were men who came looking for all sorts of metals and minerals and brought with them the financial backing to organize underground mining. This was the beginning of mining as a business in B.C., rather than as sport.

On the coast, iron deposits were found on Texada Island and the Britannia copper deposits were recorded and noted to be lacking only the financial backing for development. Most of the action was taking place on the eastern boundary of the province—in the Kootenay district, in the Rocky Mountain Trench.

One popular route to the Cariboo goldfields wound its way north along the foothills to the west of the main battlements of the Rocky Mountains. Prospectors attracted to news of silver and copper strikes in Idaho and Montana crossed the border and worked their way through the Kootenay Mountains toward the Cariboo and the persistent rumours of new gold discoveries. The occasional prospectors would hunt and peck their way up the trail and it was soon apparent that the mountains of the Kootenay district through which they passed were themselves a treasuretrove of mineral wealth.

By 1884, there were 49 claims registered around Kootenay Lake. The area had an itinerant population of almost 100 men, most of whom were Americans with hard-rock experience from places like Leadville, Colorado. These men knew that the peculiar sulfide deposits oxidizing in Kootenay outcroppings could yield lead carbonate ore enriched with silver. Silver-lead production was well advanced at Leadville, Colorado, by 1880 and the men were following the mineral trail leading north from recent discoveries at Coeur d'Alene in Idaho. In 1888, copper production began at Butte, Montana, and the entire northwest of the continent experienced the rapid development of mining and the improvement of mining technology.

The Silver King Mine was established on Toad Mountain in the Selkirk range near Nelson, B.C., in 1886 to produce silver and copper. Four years later, in the spring of 1890, one Joe Moris came to Nelson to collect money from Oliver Bourdeaux for assessment work done on Bourdeaux's Lily May claim at the mouth of Trail Creek about 40 miles south and west of Nelson. Moris learned that Bourdeaux was broke and had no money to pay him, so he went to work at Silver King just long enough to replenish his grubstake. He then headed back down Trail Creek to a claim that he had staked on his way up to Nelson. While working his Home Stake claim, where the red dirt showed through the rotting snow, Moris met and formed a partnership with another independent prospector named Joseph Bourgeois. These partners were to lay the foundations for the great Kootenay mining industry.

Joe Moris, born Joseph Maurice in Quebec in 1864, lived ninety-nine years and told his story many times. While the new partners were prospecting together on July 2nd of 1890 on Red Mountain near the Home Stake claim, they located an ore body and staked claims that were to bring great fame to the region. In one day of feverish work, they staked the War Eagle, Centre Star, Idaho and Virginia claims and Moris staked an extension to the Centre Star that he named the Le Wise. They hurried to Nelson to record their claims and were reminded that they were limited to two apiece. They offered the Le Wise extension to the deputy recorder, E.S. Topping, if he would pay the $12.50 cost of registering them all. Col. Topping (a 'Kentucky colonel') did so and renamed his own claim the Le Roi (The King)—as if in anticipation of his good fortune. He took samples of the ore to show investors in the bustling mining metropolis of Spokane, Washington, and came away with development money under the name of the Le Roi Mining and Smelting Company.

A 60-foot shaft was put down during 1891 and besides gold,

the ore was reported to run 5-20% copper with 3-10 ounces of silver per ton. The Le Roi developers realized over $86/ton for the raw ore which they shipped to the big smelter at Butte, Montana. In 1894, shipments to the Butte smelter amounted to $75,510 and the main bodies in the area had not yet been reached. During this same year, a young Ross Thompson was granted a townsite nearby and it was later incorporated in 1897 as the city of Rossland. Topping founded the city of Trail at the Trail Creek landing where the ore wagons from the Le Roi loaded their riches onto boats as the first stage of transport to the smelter. The greatest problem in the area was in finding miners to work the claims—everyone wished to be out in the hills prospecting on his own.

A business depression caused by declining silver prices crippled the mining center of Spokane and hundreds of miners there suddenly fled north to the strikes in the Kootenays where gold and copper still brought high returns. In 1895, the War Eagle and the Centre Star joined the Le Roi as showing good returns from their ore bodies. The Le Roi was becoming famous as the best gold and copper producer in the Kootenays and the other claims proved their worth as well.

The year 1894 saw the entrance into Rossland of a young American entrepreneur named Frederick Augustus Heinze. The 24-year-old "Fritz" Heinze had a degree from the Columbia University of Mines and was well versed in mining law and economics. He was already making a name for himself in the copper mining center of Montana and he sent scouts ahead to the Kootenays when he first heard of the ore bodies there. He moved into the area himself in 1895 and bought a large block of shares of the Le Roi as an inducement for that company to contract their smelting to him. With government subsidies of $1/ton, he opened his smelter on Trail Creek in February of 1897 and poured his first gold brick there in August of that year. He had a narrow-gauge railroad built to connect Rossland's mines with the Trail smelter and soon there was more ore being produced than he could handle. Other smelters were built and Heinze's organization began to waver with the competition from other smelters on the one hand, and his commitment to railroad building as a prologue to empire on the other. One person who knew him in Trail described him as being "...extremely smart, very likeable, a free spender, a real gambler and not above being crooked to serve his ends."

Heinze's smelter had been good for the Le Roi however, and shares selling in 1896 for $1.50 were soon up to $50 apiece. The mine attracted world-wide attention claiming a daily output of $84,000 worth of ore and paying dividends of $100,000 a month at the peak of its performance. The values taken from the Rossland area in the first decade of operation were reported to total nearly $5-million, and this rose during the second decade of mining there to more than $62-million.

Not far to the north of the Kootenay mining operations, the Canadian Pacific interests were busy building their railroad across the Rockies fulfilling the promise that Eastern Canada had offered the West as an inducement to join Confederation. The eastern interests behind the CP venture were attracted to the mining wealth of the Kootenays and began dabbling in various mines and smelters and services. The railroad eventually amalgamated its various holdings into one formal entity, the Boston & Montreal Consolidated Mining and Smelting Company. Shortened through popular useage to Consolidated, or CM&S, the company grew into Canada's modern national institution COMINCO.

The history of the ownership of the Kootenay mines and smelters and railroads is one of grand international intrigue in counterpoint with petty bickering, name-calling and outright scandal—the usual story of those days. In the end, Heinze's holdings were purchased by the Canadian Pacific interests, the mines themselves were repatriated by Canadian and British interests and the reduction work contracted to Canadian smelters. Heinze himself went on to further notoriety by challenging the Anaconda empire for control of U.S. copper deposits, which backfired and sparked a major financial panic and many scandals. The American interests who were slowly eased out of the Kootenays continued to use Spokane as their base and used their pay-out cash to revive that area's badly depressed economy.

Rossland in the mid-1890's was a wide open town, as reported in the Rossland Miner's 1949 history of the area "Rossland—The Golden City":

Word of the strike at the Le Roi began to spread like wildfire and to this new Eldorado fled all kinds and conditions of men. At the end of 1893 there were 99 claims staked in the vicinity; by 1895 the number had skyrocketed to 1,997... Merchants, hotel-keepers, doctors, lawyers, gamblers, painted women and all the rag-tag and bob-tail of civilization gravitated to this new strike. 'Boomers' of every description were seen coming down the hills and up the valleys.

Tents mushroomed and the scent of whip-sawn tamarac and fir was everywhere as Trail Creek basin became a seething locality. The trail that had wound to the Le Roi was widened a few hundred feet and eatinghouses, saloons, bakeries, tailors, shoe-shiners and merchants of all kinds set up for business on each side of the trail in the wierdest collection of shacks, tents and cabins ever scrambled together.

Rossland bloomed during that spring of 1893...Gold was the catalyst...Like flies to overripe fruit they came. The boa-feathered dance hall girls, the haughty and blowsy madames, the bearded lawyers, the stock manipulators and claim jumpers. Like vultures they watched the miner, jumped his claim, drank his gold, drugged him and rolled him and kicked him out into the hills for more.

And Rossland was soon to have rivals. Joe Bourgeois was back out in the hills prospecting. In June of 1892, he staked his North Star claim back up in the Purcell Mountains north and east of Rossland. The North Star was destined to become one of the largest silver producers in B.C. and its discovery produced a rush into the Fort Steele region roughly 100 miles from the Le Roi and Rossland. A quartet of young prospectors heading north from Idaho and following any news of strikes came to the site of the North Star discovery at Mark Creek within a very short time. Pat Sullivan, John Cleaver, Ed Smith and Walter Burchett found most of the hill staked and decided while they were there to investigate the hill across the creek. After two days of work, they struck the rich outcrop which was to become famous as the Sullivan Mine—the greatest lead producer in the world. They worked the property off and on for the next four years until the Le Roi from Rossland bought them out for $24,000—which seemed a fortune in those days.

In that year of 1896, the area surrounding the Sullivan and North Star claims was incorporated as the city of Kimberley. Canada's highest city at 3,662 feet above sea level, Kimberley was named to honor and to anticipate for itself the measure of wealth that was bringing fame to the Kimberley of South Africa. The combination of silver, lead and diamonds (...as drill bits) was to make the fortune of this North American Kimberley.

In the closing days of the century there were other 'booms' and 'rushes' across the face of the land. Sproule's claims at Riondel grew into the great Bluebell mine and the whole Slocan district from Kaslo to New Denver experienced a gold rush. So did the towns of Greenwood and Deadwood in the Boundary district to the west of Rossland. Here, the Canadian Pacific Railroad was a major participant. Operating their new subsidiary Columbia and Western Railroad, which they had purchased from Fritz Heinze along with his Trail smelter, the CPR extended its rails to Greenwood in 1899. When news of the gold discoveries there reached the outside world, thousands of miners and prospectors flooded the Boundary country by rail. The boom here was relatively short-lived. The day of instant wealth picking high-grade minerals off the ground was coming to a close. Large and well-organized mining companies came to town and mining became big business. Smart prospectors soon realized that there was a greater chance for wealth in locating mineral deposits to sell to the big mining firms than in staking claims they could work themselves.

The sporting aspects of prospecting and staking personal gold mines were fast disappearing everywhere in B.C. as the mining companies settled in to do business. In the Fraser district, the thousands of independent prospectors that roamed the banks and gravel bars were replaced by enormous dredging machines that gobbled their way up the middle of the river. These machines were corporate entities and all sorts of people were involved with them—one dredge is known to have been operated by Americans representing a group of enterprising farmers from Dubuque, Iowa.

It may seem that an unusually large role was played by Americans in the development of Western Canada's mining industry. This is true. With ten times the population of the Dominion of Canada and in the midst of an incredibly aggressive industrial revolution, the United States was the source for much of the technology and capital that launched the mining industry here. But it was not always such a one-sided affair. British and Canadian interests have played a significant role in the development of many aspects of the U.S. resources industry. Before the Second World War there was a fairly flexible attitude about borders and nationalities that infected the entire Northwestern region. Prospectors, miners and camp-followers floated back and forth across the border trailing the action wherever it went. The spirit of friendly rivalry and mutual assistance that marked the early competition for furs in the Northwest seems to have survived to at least a limited degree through the first half of this century. It was not so important that you were 'north' or 'south' as it was that you were a Westerner. That is what distinguished you and your way of life, and permitted the give and take that largely ignored the border.

Two Americans who played a major role in Western Canada's mining development were the Boyles Brothers. Elmore F. "Bill" Boyle was born in Iowa in 1864 and his brother Page in 1866. Their father was an itinerant railroad carpenter whom they hardly ever saw. Their mother died in 1871 and they went to stay on a farm where they worked for their room and board. In 1879, with Bill 15-years of age and Page only 13, Bill got a job for $6 per month plus room and board for them both. He worked eight months and went to school during the four winter months. After his first year of work, he bought a small house and ten acres of land in Monona, Iowa, at a tax sale for $39.00 and the boys set up housekeeping on their own.

Bill soon found a job paying $20 per month and with the stake provided by the sale of their house, the boys went West in 1886 to work at the old Silver King Mine in Arizona. They stayed long enough to earn $5,000, then moved north to Spokane and worked a number of jobs before deciding to go prospecting north of the border in British Columbia. Here they saw the trend to corporate mine development and went back to Spokane to buy a diamond drill to investigate some claims they had acquired. That first drill cost them $4,000 and they put it right to work, contracting their services as diamond drillers in the state of Washington. Their first contract was at the Republic Mine in 1895 at $2.25 per foot.

Spokane was still the supply center for the Kootenay and Boundary mining districts and the Boyles brothers worked back and forth across the border, slowly building their business. Soon, most of their work was north of the border— as far north as Alaska—and their first substantial contract came from the Granby Mine at Phoenix, B.C., in 1903. They worked Granby for many years as contractors, drilling almost 200 miles of hole there. They were to move their company to Vancouver and become one of the most prominent diamond drill contractors in Canada.

At the same time that the Boyles brothers were organizing their company in Spokane, one Fred Stone was sent by the Sullivan Machinery Company to the Silver King Mine at Nelson. The Silver King had requested a company representative to train their drilling crews for a new air-operated Sullivan drill that they were buying. This was one of the first Sullivan drills to reach the Kootenay district and when Fred saw the trend to corporate mining, he also recognized a bright future for anyone who could establish himself as a diamond drill contractor. With the added advantage that he was an old Sullivan hand, Fred made an excellent deal with Silver King to buy their new drill for himself and contract his services back to them. With his brother William and later, Ed Knight, he formed the Diamond Drill Contracting Company in 1896. Again, the base of operations was established in Spokane which was the hub of mining activity in the Northwest. Upon Fred's death in the early 1930's, Bill Stone took over as president of the company and operated the contracting service throughout the region until his own passing early in the 1940's.

The way of life in the mining camps was undergoing a massive change during this period of time, in the closing days of the Nineteenth Century. In the short space of ten years, the area changed from wilderness to the rough and raw-edged life of early Rossland to the business-like atmosphere of the mining camp described here by J. Roderick Robertson, the mine manager at Nelson, in an interview with the *Victoria Colonist* newspaper printed in November of 1900:

> In Kootenay, owing to the position of the mines, which are frequently from 4,000 to 7,500 feet above sea level, where building material is scarce and costly and fuel scanty, the conditions are as a rule not favorable for married men... Owing to the fact that fuel is costly on account of the sparse timber found at these high altitudes, and because in some

instances every foot of available timber is required for timbering in the mines...the usual practice now is for the larger mines to erect comfortable, substantial buildings capable of housing a number of men.

The cooks are important personages at the mines, receiving, according to the number of men employed, anywhere from $50 to $80 a month and "all found", i.e. no deductions for their meals...In the larger mines, there are usually one or more assistant cooks, one of whom attends chiefly to the bread baking, pies, etc. In addition, there are the waiters, or "flunkies", as the miners dub the young men who attend their wants at table. Although probably they get more chaff than "tips" the wages paid these waiters is very fair.

There is fresh meat at every meal and soup daily, in addition to which fresh and salt fish, pork and beans, eggs, oyster stews are provided. Fresh vegetables and new-made bread, biscuits, pies and puddings are supplied daily. Fresh milk in the more remote mines is unobtainable, but in its place condensed cream. On Thanksgiving Days and other high festivals hundreds of pounds of good fat turkeys and succulent hams are consumed. The management at such times frequently adding other delicacies of their own volition.

Whilst no napkins are supplied with his meals and tables are only covered with table oilcloth (although the latter is clean, owing to its being washed between each meal usually), a miner has often better food than is supplied at the manager's table, and with heat, light, food and sleeping accommodations, etc., provided for $1 per day, his necessary daily expenses are at an end. Now for clothing and luxuries. A pair of overall trousers cost him $1.25; thick underwear from a dollar or less to $1.50 apiece; a mackinaw coat from $4 to $6 according to fancy; boots from $2.50 to $4 according to the wearer's ideas. Rubber overclothing is sometimes supplied by the management in places where wet work is encountered, if not extra wages are paid to compensate therefor.

To-day the minimum rate of wage paid a miner is $3 per day for eight hours. In addition thereto in wet places and shafts from 25 cents to 50 cents per day extra is earned, and it is no unusual thing for a mine payroll to show over 50 percent of its miners receiving more than $3, for whilst the lowest paid *is $3, this is for hand drillers—machine drillers get $3.50 per day as a starter and more in deep shafts and wet places, and many mines when perched high up on the mountains 7,000 or 8,000 feet and remote from towns, pay $3.50 per day. If we take $3.25 as an average wage paid this will amount to $97.50 for a month of 30 days, or $100.75 for a month of 31 days, for miners insist on working seven days a week.*

Corporate mining became the order of the day. The Bank of Montreal took over the financially troubled Bluebell property and soon sold it to the Canadian Metal Company, a mining syndicate with French backers. Similar situations developed at many B.C. mines. Then in 1906, the Canadian Pacific interests formerly amalgamated all of its holdings in the mining world and incorporated the Consolidated Mining and Smelting Company (today's COMINCO). This consolidation of interests brought together the Trail smelters, the Rossland mines, the Sullivan Mine at Kimberley and others. With high lead and silver prices, mining was big business and CM&S imported skilled technicians and the best in mining equipment to become an international competitor in the metals markets.

Work at the Silver King Mine at Nelson began to drop off at roughly the point in time when Rossland was approaching its peak and Diamond Drill Contracting moved its crews there to drill at the Le Roi, Centre Star, War Eagle and the rest. The company continued there until the mid-1920's and extended its operations to Kimberley around 1915, shifting drillers back and forth between the towns as the need arose.

The Boyles brothers introduced a technical development to the diamond drilling world at the turn of the century that revolutionized the industry. They were the first people to devise the double-tube core barrel.

At the time of the First World War, the development of the process of differential flotation made it possible to exploit complex ore bodies which had until that time been considered worthless because there was no way to separate the ore into pure metals. Low-grade ores could also be made profitable by separating the various metallic elements into high-grade concentrates. In western Canada, this led to the formation of the fabulous zinc industry.

In 1919, CM&S came up with a flotation system to separate the complex sulfides of the Sullivan orebody. A concentrator was in operation by 1922 and Kimberley suddenly became the center of the Kootenay mining scene. The Sullivan Mine continues in operation today and has produced one hundred and fifteen million tons of zinc-lead ore plus smaller amounts of iron and silver.

Copper was the major product of the Britannia mines on the coast just north of Vancouver, gold-silver the item that triggered a mad rush into the rugged country of Portland Canal on the B.C./Alaska border. The Anyox mines at Observatory Inlet opened in 1910 and there was action in the province's coal deposits—almost all of the activity developing from strong corporate bases. Some companies dabbled at re-opening older properties, but for the most part the mining industry in B.C. spent the first quarter of the 20th Century working existing properties for a steady return.

II

In the mid-1920's, certain fundamental changes took place which established the configuration of the modern diamond drilling industry in western Canada. The ownership of the Boyles Bros. firm changed hands and the T. Connors Diamond Drilling Company was founded. These two contractors have played the major roles in the history of the industry ever since that time.

James A. "Jim" Cullinane was one of those present at the founding of the Connors firm. He was of the first generation to be born in the mining camps of the Kootenays where his father was one of the most respected percussion drillers at the Centre Star Mine in Rossland. Jim's generation supplied the first wave of native-born talent to the diamond drilling industry, including Norman Allen of Rossland, C.A. "Chet" Bartoo and Al Morehead of Spokane and Eddy Nord of Kimberley. Jim continued with Connors until his retirement in the early 1970's and his son, James, is one of the modern company's executives.

Jim Cullinane is a storyteller and his recollections of that period in the industry's history bear repeating. They reveal more about the life of the times and the working conditions of the drillers and their personalities than all the written histories combined:

> "My father had some ideas that I should become a mining engineer, rather than a miner like he was himself, so I came down to Vancouver in 1924 to attend classes at the University of British Columbia. I went back home to Rossland at the end of the next spring looking for 'summer dollars'. There was plenty of work still going on up the hill, so that's where I went looking for a job.
>
> "It was one of those Monday mornings when the crews

were all split up and a helper could fit in just about anywhere. It was well into the spring and everyone was working underground. I went up to the drill shack, which was called the setting shack in those days because that's where the foreman restored the drill bits, and there I was to talk with Tom Connors.

"Tom Connors came to Rossland in the early 1900's to work as diamond setter and general foreman for Fred Stone's Diamond Drill Contracting Company. A great many of the early drillers in the Kootenays came from the U.S., from work in the Mesabi—the 'Iron Range'—in Minnesota. A good proportion of these men were Scandinavian and came to be nicknamed 'Michigan Swedes' coming out of that region in the U.S. Among those coming up to the Kootenays besides Tom Connors were Ole Karlsson, John Kier, Jim Cohen and the brothers Oroville and Howard Yarborough. Just after the First World War, drillers began drifting west from eastern Canada: Ezra Campbell came from Newfoundland, Ade McClelland also came west from the Maritimes, Colin McCutcheon, who fought with the Seaforth Highlanders, came west after first working at Sudbury, Ontario, Frank Hunt came over from England and Mike Donaldson came through eastern Canada from the States. It was this group of men who gave diamond drilling its boost in the Kootenays.

"At the time I went up to see Tom Connors, he was foreman in Rossland for DDC which was headquartered in Spokane, and that made him superintendent of the entire operation. He was perched up on his setting stool working at the setting bench and there were three other men standing around the shack.

"I was a gangling green kid—I'd been underground before, of course, because my dad was a miner and I grew up there in Rossland—but at my age everyone over thirty looked like an old man to me and I paid deference to them all. Tom pointed at me from his bench—he did all his talking from the top of that stool, you see—and he said: 'Here's a new helper, you better take this lad with you'. The man he pointed toward, Mike Donaldson, said: 'No Tom, I'm moving that Sullivan C machine and that's no job for a kid'. And that was that. The old-timers were all independent and they had quite a say in what went on, back then.

"There was another fellow there, a Cockney Englishman named Frank Hunt with an accent to make you think he'd come over the week before last. Frank had two or three brothers who were already quite well-established in business in the Kootenays and while they all had some accent, the others were nothing like him. He cultivated it. He had maybe four full fingers in total, the rest were pretty chopped from getting caught in the open gears of the rigs. When Tom looked at him, Frank said: 'Naow, Taowm, I'm workin' in that deep 'ole down thair and that's now place for a green 'elper'.

"That's two of them let me down and I'm thinking I should sneak out between somebody's legs, get out of there and try my luck at something besides diamond drilling. The other man standing in the setting shack was Ole Karlsson, an old Swede, who always had a little stubble beard. He was dressed in his big rubber boots puffing on a brown paper cigarette. 'Come on kid,' he said, 'you come down with me.'

"I went on down to the 1200-foot level of the Le Roi Mine with him and while I'd seen a diamond drill at work before, I'd never had a hand on one. He was working a flat horizontal hole and he was in 1800 feet. All the drill rods were out of the hole just then and stacked in the shaft. It just looked like a couple of miles of plumbing pipe to me. But Ole would teach you to work, to keep busy all of the time. As long as you worked every minute of your shift, he'd bend over backwards for you.

"In early summer, they started work up on the mountainside on the surface not far from my home. By this time there had been some layoffs and since a drill helper had to put in a year or more before he learned enough to be a driller, the layoffs scared us younger men a bit. The fellow cross-shifting me had been a helper for a couple of years on and off. I met him in town one day and he told me he'd been let go and that I'd probably get my notice that afternoon. I went home and told my mother not to pack a lunchpail. She gave me one anyway saying you never knew what might happen at the mines. I went to work as usual and the Swede didn't say anything, nor the boss, so I kept my mouth shut. This went on for three or four days.

"Then I had to go down to the shack for a new bit and the bitsetter there was Bill Nicolls. The set-up shack was just four poles and a roof. He asked me: 'You ever stop to wonder in the last week or so, how come you're still working when so many others have gone?' I said: 'Yes, I did wonder, but I figured it wasn't up to me to ask questions.' 'Well, it's because of that cranky old Swede! When he heard that the layoffs were coming he told us he was keeping HIS helper—just like that!'

"I continued working with Ole Karlsson until September while I was getting ready to go back to school. Then Bill Nicolls asked me why I was heading off so soon after I had some good experience behind me and made me an offer to stay. And so I stayed for almost fifty years!

"I worked on the surface with Ole until the middle of 1925. Then DDC put out a call for drillers to go out on a job to a remote area north of Edmonton—on the border of Alberta and the Northwest Territories. This turned out to be the last exploration job that DDC contracted in Western Canada. It was all put together in the ten days before that Christmas and I signed on. We left New Year's for the north.

"There were six of us Canadians and the remainder of the crew was from the Spokane area. We met in Edmonton and went north on Alberta Great Waterways Railroad to Fort McMurray. The drilling crews were supervised by Chet Bartoo, the others in the party being exploration engineers from the Mining Exploration Company (Victoria Syndicate) led by H.L. Batten. We mobilized in Fort McMurray with two complete drills, drilling supplies, foodstuffs and camp gear to last 35 men until the breakup in the spring. All of this was loaded on horse-drawn toboggans—broken down and sized and weighed so that each load was roughly equal. We each got a horse and toboggan to look after under the supervision of the Ryan Brothers Transportation company of Fort McMurray.

"We began to leave town the next day, early in January of 1926, with eight to ten sleds in a shift. It was necessary to travel in shifts so there would be adequate sleeping room and shelter for the horses in the wilderness cabins along our route.

"Many of us had never experienced the intense cold of the north, but day after day we trekked through cold of -30°F. and below. It was quite an experience for a young man like myself. We'd make perhaps 12 miles a day—sometimes as much as 20 miles, traveling on the river ice.

"After 14 days on the river, we reached our camp. A

crew of local Indians had just finished the construction of the cabins begun three weeks earlier. Green logs were cut in sub-zero temperatures, the cabins built and roofed with split jackpine poles. Once we started heating those cabins, it drew the frost out of the logs with the result that we had all white, ice interiors. All in all it was quite a winter. Through some errors in food estimates, many of our meals left a good bit to be desired. And we had to become instant experts in core drilling in sub-zero conditions. We were fortunate that the rock formation being drilled was good coring rock and we made relatively fast progress with the job.

"In the end, it appeared that someone down south had gotten a bum steer on the mineral finds in the area and the exploration came to nothing. By the time the first boat came upriver after breakup, we were all ready to leave. Half the group went out with that boat and the rest went out on the next two boats to come along.

"I got back to Rossland from the north at the end of June and the first thing I did was to go up the hill to see Tom Connors. Connors Drilling had been incorporated while I was gone. Fred Stone had left the area to go to California to team up with a fellow named Doheny in the oil-well drilling business, which was still mostly cable-tool work. His brother, Bill Stone, and Ed Knight were running most of DDC's work down at the Republic Mines in Washington State. Tom Connors was superintendent in Rossland and Pete Murphy was superintendent in Kimberley.

"P.D. 'Pete' Murphy was a great character. He was drilling at Nelson in the very early days of DDC and then went to Rossland when the action built up there. After the Sullivan deposit was producing, he was moved to Kimberley to run the DDC branch there. That was in the early 1920's.

"Now, in the early years of diamond drilling and extending into the early 1930's, all diamond bits were hand set, using black carbon stones instead of the boart or industrial grade stones of today. Drilling crews were subjected to extensive training on the proper use of the bit, because it was so much more valuable than it is today. Good quality stones sold for upwards of $100 per carat.

"I was the goal of the driller to become a diamond setter. The setter was, in most cases, the actual foreman of a project. On big multi-drill projects several setters might be employed under one general superintendent such as Tom Connors or Pete Murphy.

"Drilling at Kimberley presented one very great problem. Much of the underground exploration drilling had to penetrate the famous Kimberley granite/chert formation, known as one of the hardest rocks in the world. Pete Murphy was a superb bit setter and he made the Kimberley job his own personal operation by coming up with a drill bit that the other old-time bit setters just could not duplicate. It was known as a fragment bit. Before Pete came along, the drillers there used 'chip' bits which were quite small—say 20 to a carat. Many a fine driller flattened this out in 5 to 10 feet of drilling and the bit would have to be withdrawn from the hole and reset. Pete's bit consisted of fragments of black diamond that were in reality just tiny slivers of the stone. His bits would get 20 feet and sometimes as much as 30 feet through the chert before they would have to be reset.

"For perhaps a year, Tom Connors at Rossland and Pete Murphy at Kimberley had run their DDC operations as autonomous branches. In the winter of

1925-26, they got together with the CM&S manager of mines, Mr. Archibald, to talk over future exploration prospects for that company. It was decided to split up the Canadian contracting company and the American parent company. With $30,000 of borrowed capital, they formed the T. Connors Diamond Drilling Company Ltd. and established their headquarters at Kimberley. The original shareholders included Tom, Pete, Bill Stone, G.C. 'Glen' Marshall and Norman Allan. Within a short time, Jock Patterson, John Soppitt and myself bought shares in the new company.

"In the spring of 1926, Connors Drilling took over the CM&S contracts from DDC and bought all the DDC equipment on the Canadian side of the border. This was the state of affairs when I returned from the north. Tom Connors told me that Pete was organizing work on the phosphate deposits up in the Crowsnest Pass in the Rockies and asked me to wait around a bit. I took a few days off, then went to work with Tom. When Pete called, he told me the job was a bit unusual, something new to the area—drilling in shales. Tom asked me if I could make it go as a driller. I said I'd try my best. He snorted: 'Your best ain't good enough!' And I was off for Fernie with my first big assignment as a driller for DDC in the North.

"The next winter, we lost Tom to pneumonia on January 27, 1927. He had traveled to Kimberley for meetings with Pete and to socialize with our major clients attending a curling bonspiel there—the Kootenays were the center of the sport at that time. He took sick and failed so suddenly that he was dead at the age of 45 years before his wife could travel up from Rossland to nurse him.

"Pete Murphy ran the company alone for the next two years. It was a tremendously difficult job because he acted as president, general manager, office manager and field supervisor all at the same time. In the early part of 1929, he acquired the services of D.C. "Dunc" Chisholm from CM&S to operate the business end of the company. Dunc eventually became president of our company, a post he held until his retirement in the late 1960's.

"Now in the early days, practically all the company contracts were with CM&S and this left the impression with some people that we were a part of that firm, which was never the case. It was simply that Tom and Pete had worked for so long as their contractors that Connors could anticipate the needs of that particular client and cater to them.

"When the Rossland mines began to close their doors during 1928-29, the work shifted to the Sullivan mine at Kimberley. At the same time, exploration drilling jobs began to develop away from the Kootenays. Several jobs were undertaken by Connors on the Pacific coast. In 1929, we began a surface drilling program in northern Saskatchewan in the area of Lac La Ronge and Rottenstone Lake. The equipment for that job was freighted in by tractor train from Prince Albert, but the crews and camp supplies came by aircraft. The planes used for this work were mostly World War I Junkers and Norsemen.

Shortly thereafter, we began drilling near Yellowknife in the Northwest Territories and again used the airplane for support.

"Some overseas work developed as well. Back in 1925, when I headed north, Oroville and Howard Yarborough took a crew of Kootenay drillers to Spain to drill for DDC. In 1928, the Sullivan Machinery Company itself recruited Kootenay drillers for work in Africa, in Rhodesia, at the rate of $300/month plus all expenses.

Right through the 1930's, various companies were always recruiting Kootenay drillers for work around the world.

"I had to forego my trip to Africa to work an exploration program on various properties in the Portland Canal area north of Stewart on the British Columbia/Alaska border. Working AT&T and George Copper properties the terrain was mostly vertical. All the equipment had to be packed by humans and horses over very steep mountain terrain—there were no roads. Looking back from this day of workhorse helicopters and pre-fabricated camps, it all seems so very unreal. Those days float through memory like some kind of dream."

At the close of the 1920's, the Diamond Drill Contracting Company remained active in the United States. The company established headquarters in Coeur d'Alene to be near the work in the Idaho silver mines, then moved back to Spokane to be close to airport facilities. For a number of years, DDC and Connors used to trade drilling personnel back and forth across the border and Jim Cullinane was just one Canadian driller who saw work at the silver mines in Idaho and on dam construction in Montana.

Practically all the drills of this era were made by the Sullivan Machinery Company. Underground, they were powered by compressed air from above. On the surface, they were powered directly by steam. At the end of the 1920's, there was a revolution in diamond drilling technology which had its roots in the Northwest. The contracting companies working here recognized the need for small, light-weight drills which could be easily moved over rough terrain. The Mitchell Diamond Drilling Company of Spokane developed and manufactured the 'Light Mitchell' drill and a compatible line of drill rods and supplies.

Ed Knight and Bill Stone, continuing to work as DDC

contractors, developed a skid-mounted drill powered with a gasoline engine. At first they used a 2-cylinder Cadillac engine for power, but changed to the Ford Model T engine because of its simplicity and reliability. Their drill had a side-mounted swivel head instead of one mounted at the end of the frame to facilitate horizontal drilling. This 'Sidewinder' drill became the constant companion of Connors drillers and a few were still being used in the late 1940's.

Important as they were in signalling a change in drilling technology, the 'Light Mitchell' and 'Sidewinder' drills were overshadowed by the technical achievements of the Boyles company.

Following the death of his brother Page in 1917, Bill Boyles continued to operate the very successful contracting company in partnership with two Scandinavians who were very active in the Pacific Northwest, Laurids Lewed Jessen—a Dane, and Frederic E. Lindhe—a Swede. The business was moved from Spokane to Vancouver, following changes in the pattern of diamond drill work as it shifted north toward northern B.C. and Alaska. In 1926, Jessen and Lindhe bought the business from Bill Boyles who was in poor health. He died in Vancouver at the age of 64 years.

Jessen and Lindhe soon bought the Dan Longton company in Salt Lake City, Utah, to re-establish their American interests. Both men were primarily interested in the business of contract diamond drilling, but Jessen began to develop an interest in the technical aspects of the business as they related to the manufacturing process. He began to experiment with designs for compact, light-weight drills in response to the demand for a portable unit for use in the bush in western Canada. His first drill was a 1,400-lb. unit that could drill cores of approximately 1½" diameter to 1,500 feet. Powered with a Model T Ford engine mounted on the forward section of a Model T frame—radiator and all—the gas-powered drill quickly took the place of the unwieldy steam drills in widespread use at the time. The swivel head and hoist were the same Longyear parts used on the steam drills, but the Jessen rig was skid-mounted and could be moved around in the bush using its own winch and cable for motive power.

At the same time, in 1929, he built a small drill for use in prospecting in the bush. This was the famed X-RAY drill which weighed only 140 lbs. and could be backpacked into remote areas. It could drill a 3/4" hole to 150 feet. Within two years, he developed a light-weight underground drill powered by a 2-cylinder double acting 'V' compressed-air motor, which weighed only 300 lbs. complete. With a capacity of nearly 500 feet, it recovered a core of approximately 1" diameter.

Beginning around 1928, the mining community in western Canada began to publicize its love affair with the airplane. The newsmagazines were filled with articles over the next few years extolling the virtues of aircraft as the major method of bush transport for prospectors and exploration teams. Eventually the fever spread to include mention of just about every type of aircraft imaginable. The July 1929 issue of the *B.C. Miner* magazine features an article, for example, arguing the merits of various "Buoying Gases For Airships".

The 1931 introduction of the production model of Jessen's X-RAY drill was widely publicized as a machine designed for air transport. This had an immediate effect on the exploration industry and on the fortunes of the Boyles company. In the opening days of the Depression, Boyles Bros. Ltd, was able to open new production facilities in Vancouver which prospered in the darkest of times. Jessen capitalized on the success of the X-RAY by introducing the BBS-1, a 550-lb. machine with a capacity of almost 650 feet of 1" core, which was highly portable. In years to come, the BBS-1 became the workhorse of the exploration drilling contractors. The BBS-2 replaced Jessen's very first skid-mounted drill as the machine for long-hole drilling. These machines used Ford engines, but the other parts—including drill heads and winches—were of Boyles manufacture.

Exploration technology took several giant steps forward during the 1930's. The September 1932 issue of the *B.C. Miner* reported that a young assistant professor from the physics department at U.B.C., Gordon Shrum, announced the invention of a portable field instrument for detecting radioactive ores. The invention of Shrum and Ronald Smith, Dr. Hebb and Dr. Seyer was designed as a backpack with a thin pencil-like probe that could be dropped down a diamond drill hole to detect the presence of radioactive ores. This was one of the first field applications for the Geiger-Muller device.

Another major landmark in diamond drilling technology was the cast set drill bit. In the early 30's, a number of suppliers and one or two of the contracting firms experimented with a cast set bit. Black diamonds had long since advanced in price to the point of being prohibitive for use in drill bits, but the small boart stones were very tedious— and thus very expensive—to set by hand. The cast set bit was found to be very adaptable to the boart diamonds. By the mid 1930's, Boyles Bros. Ltd. was advertising its 'Ready-Set' bits at $35 apiece.

Toward the end of the decade, T. Connors Diamond Drilling Company made a deal with Process Diamond Bits of San Francisco to manufacture on a royalty basis their design of a cast set bit. John "Jock" Patterson, who had joined Connors in Kimberley in 1928 as a drill helper, was sent to the U.S. to learn the process. Upon his return from San Francisco, a bit setting facility was opened in the Connors shop in Vancouver. Under Jock's supervision, Connors was soon producing its own competitive bit and the company bought the process outright from the American firm.

The Boyles company, with its new drill designs and prosperous manufacturing arm, was in no danger of financial disaster during the Depression years, but this was small comfort to Jessen who had a conception of his firm as a major international force in the field of diamond drilling. He spent the Depression years seeking out new markets for Boyles to exploit. In the May 1933 issue of the *B.C. Miner*, he made a strong case for the increased use of diamond drilling as a means of developing existent mining properties. He suggested that the diamond drill should be imagined as the mine manager's 'x-ray' to follow ore veins and to plot new work sites. He mentioned in his article that Canadian drillers were traveling around the globe promoting the use of diamond drills—and he would be the one to know about that, because most of them were Boyles men working jobs that Jessen himself developed.

In 1932, Boyles established a base in eastern Canada at Kirkland Lake, Ontario, and in 1935 they opened another office at Port Arthur at the Lakehead. That same year, Boyles established a contracting branch at Manila in the Phillipines and opened a sales office in Johannesburg, South Africa-- just down the road from the boart diamond people. The company acquired the Alberta Diamond Drilling Company and set up another contracting branch at Kitwe in Northern Rhodesia.

The year 1935 saw a split in the Boyles management when Fred Lindhe purchased the American contracting operation from Jessen and incorporated the Boyles Bros. Drilling Company of Salt Lake City, Utah, with his brother Charles. The next year, Boyles Bros. Ltd. was purchased from Jessen by a number of the company's younger executives who changed the name to the Boyles Bros. Drilling Company Ltd.—which had no relation whatever to the American firm. Two years later, Boyles of Canada acquired Diamond Core Drilling Pty. Ltd. of Australia and incorporated yet another branch in the Orient at Singapore. Contract work was also carried out in East Africa where blasthole drilling was done on copper properties in both Northern and Southern Rhodesia.

155

Jess L. Havlick, a Kootenay driller who worked for Pete Murphy and DDC in Kimberley in earlier days, went to Africa to manage the Boyles Bros. Drilling Company (Africa) Ltd. In 1933, he was joined there by Arthur E. "Shorty" Hammerlund. Hammerlund bought that company for himself the next year, and Jess Havlick went on to Australia to play a major role in building the Boyles subsidiary there. Other Kootenay drillers traveled in the 30's to South America, Central America and the U.S.S.R.

Other diamond drilling firms in the west met the problems of the Depression years with different degrees of success. Ken Robinson Diamond Drilling of Vancouver did a good bit of work on the coast and landed the contract to survey the right of way for the Pacific Great Eastern Railway. The Vancouver Island Diamond Drilling Company, owned by Petrie, kept reasonably busy and Petrie himself went on to fame with his re-design of a German invention into the first practical IEL chain-saw, which found wide acceptance in the B.C. forest industry.

The original Diamond Drill Contracting Company of Spokane halted its operations in the mid-30's in response to the adverse economic conditions just south of the border, but after a few dormant years, the remnants of the company were purchased by William Burrows. DDC was rebuilt under the original name and remains active today under the direction of William's son, Lees Burrows.

Connors Drilling was one company which was forced to cut back on its operations during these lean times. While a small group of old-timers continued to operate the company's long-term contracts, some of the younger men like Jim Cullinane went their own way in search of survival money. Jim joined his father and some friends to reopen workings at the old Ymir Mine north of Nelson, B.C. Pete Murphy had worked Ymir in the very early days of DDC and there was a bit of irony in reworking that old property. But there was a trend toward reopening old mines at the time—gold is a 'hard times' metal and it was particularly attractive in the early 30's. Most mineral deposits of economic significance were corporate properties by that time, except the old placer gold mines. Small deposits that were considered exhausted by corporate standards could still be worked for a fair profit by small independents. Hired help went for $4.50 per day during those years and gold advanced in price above $20.67 a fine ounce. By 1933, the U.S. dollar was so far ahead of the Canadian dollar in value that gold was sold through the Canadian mine to the U.S. for almost $35 an ounce. Jim's group made sufficient funds during the better nine months of the year to take off the three severe winter months and go curling.

Those who stayed with the company worked many odd jobs in remote areas. Barkerville and Hedley were once again the scenes of some activity and the Bridge River area was just coming into its own. In 1929, the Pioneer Mine was brought into profitable production and two years later, Bralorne Mines Ltd. was formed from the old Lorne and Bradian properties. Connors drillers worked regularly at these remote sites in the Coast Mountains. By the mid-30's, CM&S contracts were taking Connors drillers to Nova Scotia and northern Ontario. When this work was done, Pete Murphy developed additional jobs in the East and Connors opened a temporary office at St. Lambert near Montreal to supervise work on the Shipshaw Dam at Arvida, Quebec.,

An extensive drilling program was carried out at Goldfields, Saskatchewan on Lake Athabasca. It was here that the company acquired the services of John Elneff. Known to his friends as "Jackfish Johnny", Elneff had a colorful history in northern Alberta and the Northwest Territories. He arrived in Canada at the beginning of the Depression from his native Denmark. With no work in sight in the populated areas of the country, he headed 'down north' from Edmonton into the Arctic. His first job was on a supply boat on the Mackenzie River which was trying to make one last trip before freeze-up.

They made it down the Mackenzie, but not back up again and Johnny had the misfortune to spend eleven months of his first year in Canada frozen into the Arctic ice. Afterwards, he got a job cutting wood for the Athabasca river boats, he fished, he traded and then he took off for the goldfields.

With a boat and an outboard motor, he ran a water taxi service on Lake Athabasca during the summer. When winter came, he bought a Model A Ford and ran taxi over the ice. Always seeking a challenge, he joined a Connors crew working in the vicinity and became hooked on diamond drilling. He was soon respected as one of the most able field supervisors and job foremen in the industry. Retired now in Richmond, B.C., he was especially skilled at organizing and supervising 'bush' jobs and his skills contributed a great deal to the survival of the company in hard times.

Those who remained on the West Coast moved the Connors head office from Kimberley to Vancouver in 1933. The company's work was spreading so far afield geographically throughout Canada that a more accessible urban location was desirable. Many of the major mining companies who were Connors clients maintained offices on the coast and communications were improved with the move to an office located downtown on Hastings Street.

III

World War Two brought major changes to the diamond drilling industry in Canada. With the onset of the war in the Pacific, Boyles lost its subsidiaries in the Phillipines and Singapore, and some of the company personnel were interred in prison camps. The Australian branch became something of an albatross around the company's neck, because war-time restrictions forbade the removal of money from the country and the Canadian firm could not continue to support the Australian branch without some return. A deal was arranged with Kirkwood and the contracting branch sold, as well as certain manufacturing rights to allow Kirkwood to manufacture Boyles products under a royalty agreement. The new manufacturing plant was reported to be almost identical with the Vancouver installation. Today, the Kirkwood firm is one of the world's largest diamond drill firms and is known as Mindrill.

At home, in Canada, many of the younger drillers reported for induction into the army. Those drillers left behind were swamped with work. The Canadian government asked gold miners to produce as much as possible at any cost to boost Canada's foreign exchange potential in expectation that Canada would presumably be the chief purchasing agent for England for arms and munitions from the United States. Many of the gold mines were high-graded to provide maximum values at considerable development cost.

Then, with the U.S. entry into the war, Ottawa announced that other means had been found to meet that particular exchange problem and the gold mines were stripped of their manpower to benefit the base metal producers. The emphasis quickly shifted to strategic metals since many of the world's major production centers were inaccessible to the Allies. Western Canadian mining companies explored and developed new sources of mercury, tungsten and other metals in this country.

CM&S brought in a fabulous mercury mill operation at Pinchi Lake and Bralorne another at Takla Lake, both in northern British Columbia. Connors Drilling was responsible for the exploration drilling at Pinchi Lake, just north of Fort St. James and John Soppitt supervised the job. The job involved tremendous pressure from authorities for a new mercury supply and while the exploration drilling was underway, a mill was already being built on the shore of the lake. In every sense of the word, Pinchi Lake was an 'instant' mine. Tungsten was another strategic material and mines were established at Red Rose and on the old Emerald Mine property.

Jim Cullinane moved back to Nelson from Vancouver to supervise Connors work in the Kootenays. Six miles out of Nelson, there was additional development of the Kenville Gold Mine. The Hedley nickel properties were working again and in 1942, CM&S began to explore the Bluebell one more time. It was still considered unworkable, but five years later the extraction problems were solved and the mine went directly from exploration to production and Connors remained the exploration contractor there from 1947 until the early 1970's when the mine ceased production.

For the drillers who remained in Canada, there was plenty of work to be done and for those who went into the army, there were developments which had a bearing on the post-war industry. Canadian miners played a significant role in the First World War, mining and counter-mining the major battlefields of the Western Front. After Munich, with the threat of another world war becoming apparent, the Canadian mining fraternity made a proposal to the government that Canadian mining engineers tour those parts of Europe which might be enveloped by the war in an effort to anticipate the technical problems the western armies might face in breaking the German Siegfried Line. It was suggested that the new mining technology developed in Canada since 1920, especially in the realm of drilling technology, might be of value in this endeavor.

A meeting was held between military men and representatives of the mining industry to discuss the feasibility of using diamond drills to drill long horizontal holes under Germany's permanent fortifications for the placing of explosive charges. General A.G.L. McNaughton of the First Canadian Division proposed that Ontario's former Minister of Public Works, now Lt.-Col. C.A. Campbell, assemble a unit of hard-rock miners and their equipment and proceed to England to begin experimenting there with diamond drilling for use as the handmaiden of demolition. The Special Detachment was put under the command of Lt.-Col. C.S.L. Hertzberg who was later to become Chief Engineer of the First Canadian Army. Perry Hall, Boyles' field superintendent for Eastern Canada, was attached to the unit as Civilian Technical Advisor to act as the liaison officer between the Canadian War Time Mining Association and the special diamond drilling section attached to the special unit. He was subsequently commissioned a Captain in the Royal Canadian Engineers and was promoted to Major by the war's end. Perry Hall was an ingenious man with an intuitive grasp of technical problems. He had a great deal to do with the design or modification of the special equipment required for use by the overseas tunneling companies during the war. In years to come, he would remain the major influence upon the growth and prosperity of the Boyles company, and later become the company's president.

The work of the diamond drill unit seized the imagination of the British high-command, and plans were developed to send the unit to France to prepare signal cable holes that would be invisible to air observers. When the unit was packed and ready for deployment, the Germans overwhelmed France in their incredible push to the sea. The Canadian drillers unpacked and resumed work in England, helping the British re-open mining properties of strategic importance. The Boyles company became a prominent fixture in England during the war.

Just before the Japanese invaded the Phillipines, a Boyles driller named Smith was sent to Nigeria with two Filipino helpers to conduct exploration drilling. The job over, it was impossible to return to the Phillipines, so Smith headed for England and put his drill to work in the coal mines of Wales. In effect, Smith started a Boyles contract drilling operation in England which was soon expanded by the company. Headquartered at Newcastle-upon-Tyne, the new branch acted as the liaison between the Vancouver-based manufacturer and the English government and the Canadian overseas detachments, supplying diamond drill equipment and services to wartime England. After the war, a manufacturing facility was built there which remains in existence.

The No. 1 Tunnelling Company in England was charged with the job of mining the runways of all the airstrips in the British Isles to prevent a surprise air invasion by the German Luftwaffe. This job required the drilling of long horizontal holes in a series beneath the runways, which were packed with explosives. A heavy-duty machine was required with the type of side-mounted swivel-head which characterized the Knight & Stone 'Sidewinder'. Boyles Bros. Ltd. modified their BBS-2 for this use. Boyles had the contract to supply all the drills for the war effort and the Longyear company supplied all the auxiliary drilling equipment. After the war, this modified BBS-2 drill found commercial success as the Boyles 'Slidemaster' and was used primarily by highways crews to drill drain holes and thus help stabilized large rock masses on slopes above Canada's highways and rail lines.

In the latter part of October 1940, an urgent request was received in England that a detachment of hardrock miners be made available to help construct defensive works of considerable importance at Gibraltar. The Canadian Special Detachment of about 100 men under the command of Major Campbell arrived at the Rock on November 26th, 1940.

The great underground fortifications of Gibraltar were

begun during the Great Siege of 1779-83 when the French and the Spanish tried to recover the real estate from the British garrison under Sir George Elliott. The fortifications were built under the direction of Sergeant-Major Ince of the Company of Soldier Artificers, forerunner of the Royal Engineers. Little more was done until the 20th Century when technical advances in tools and explosives made it possible to tunnel into the heart of the Rock. Legend greatly exaggerated the extent of the tunnelling there and at the start of the Second World War, the British realized that there was a fair bit of work ahead to make the Rock impregnable to the modern war machine.

The detachment of Canadians from the No.1 Tunnelling Company had scarcely begun working when Canada was asked to provide additional help on this front. The Canadian miners were quickly recognized as the world's experts at hardrock mining. The No.2 Tunnelling Company was formed under the command of Major C.B. North of Vancouver and sent to Gibraltar in March of 1941. When they arrived, a part of the first Special Detachment returned to England to supervise drilling there. The Canadians were not the only miners on the Rock—the British had a number of tunnelling companies there—but it was the Canadian group that introduced the others to diamond drilling and taught them techniques for using the new machines coming out from Vancouver. In May 1942, command of the unit passed to Major J.G. Tatham of Larder Lake.

The chief task for the Canadian drillers and miners was the construction of a massive underground hospital that would be secure from any enemy attack. Work began on the excavation in March of 1941 that was to be named the Gort Hospital after Field-Marshal the Viscount Gort. The drillers followed the deposit of dense gray limestone through the interior of the Rock in advance of the mining detail. For two years the miners worked to turn the Rock into a network of wards, operating rooms, storage compartments and living quarters.

The wards were the largest chambers, approximately two hundred feet long by thirty-five feet in width with twelve-foot ceilings. The Rock of Gibraltar remains one of the most remarkable military hospitals in history.

Running through the interior of the Rock and connecting the various parts of the hospital was a single major tunnel twelve hundred feet long by twelve feet wide with twelve-foot ceilings. Named Harley Street after the famous street in London where the medical profession was concentrated, it ended in a massive laundry room where the tunnellers scraped out eleven thousand tons of rock and debris. There was other work on the Rock as well, including the excavation of ammunition dumps and fuel storage tanks and the construction of blockhouses and fortifications.

In their two years on the job, the Canadians expended 46,000 man hours on construction and removed more than one hundred forty thousand tons of rock from the interior of Gibraltar. In one week in 1942, they averaged over four tons per man per shift. On December 5th of 1942, the last shift was worked underground and within ten days the Canadians were headed back to England.

There was a second group of Canadian drillers on the Rock at the time. Shortly after the return to England of some of the members of the very first Special Detachment, the Canadian high-command assembled yet another group of diamond drillers for secret work at Gibraltar. The second Special Detachment was formed under the command of Capt. H.W. Demorest, proprietor of one of eastern Canada's largest diamond drill contracting firms, Demorest Drilling Ltd. of Noranda, Quebec. At war's end, the Demorest firm merged with Boyles Bros.

The twenty-five men of the second Special Detachment arrived back at Gibraltar in February of 1942 and set to work enlarging the airport facilities there, extending the runway into the sea using the scree from the base of the Rock. The scree was too large for the construction equipment the Canadians had at their disposal, so Capt. Demorest adopted the method of using high-pressure water to bring down boulders from the side of the Rock. After 'hydraulicing' the material down, the Canadians broke up the scree with diamond drills and explosives.

During the early days of November 1942, the Canadians watched as Gibraltar became an important staging area for the Allied offensive in the Mediterranean. On the 8th of November, word was received that the great armada of ships and planes that used Gibraltar as a base had landed in Algeria and in Morocco to open the offensive in North Africa.

Many more Canadians were to see Gibraltar in their uniforms as they funnelled into the Mediterranean headed for Africa and Sicily and the Italian peninsula, but none would know it so intimately as the Canadians who helped 'build' the Rock.

And then the war was over. Everyone returned home to find that the world had changed, sometimes for the better and sometimes not. The contracts held by Canadian mines with U.S. government purchasing agencies expired and the mining industry found itself left high and dry for a time. The production of metals in B.C. dropped from a 1943 high of $46.5 million to $35.8 million during 1945.

The incentive to expand and open new markets which often follows times of war was missing for much of the Canadian mining industry because a return to peace meant a return to the restrictive taxation policies instituted to cope with the Depression years. It took some time before the industry fully recovered from the extraordinary demands of the war planners.

Typical of the post-war activity was the acquisition of the

old Emerald Mine at Salmo from Wartime Metals by Canadian Explorations, a Placer affiliate. The primary production was tungsten which was still in short supply. Mine manager H.L. Batten, of the old Victoria Syndicate, brought the Emerald into lead-zinc production and for two decades the old mine remained a good producer.

Diamond drill technology made a few small advances in the immediate post-war years. Throughout the 1940's, several diamond bit companies experimented with bits using a hard metal matrix that appeared to be superior to the soft matrix, especially in drilling broken or friable formations. The Connors company looked at several different bits using the hard metal matrix and found one with great potential. After considerable field testing, a deal was made with J.K. Smit & Sons of Toronto to produce this bit as a joint endeavor. A new company was formed, Smit-Connors Ltd., and a plant established in Red Deer, Alberta.

Jock Patterson left Connors to become the manager of the plant because he was widely respected as a fine craftsman. In a few years time, Connors sold its interest in the company to its partner and it was absorbed into the family of Smit operations. Jock Patterson continued as manager in Red Deer until his retirement in 1971.

Having proved itself of such value during the war for blast hole work, diamond drilling was once again considered for such work in the peace-time industry. At this point in time, the technical superiority of the diamond drill was displacing many of the older methods of mine drilling. This was to change again within the decade with the introduction of more sophisticated pneumatic products, but in the late 1940's it was successfully argued that diamond drills were ideal for blast-hole drilling. C.H. Hopper argued well in the *Northern Miner* in mid-December of 1948 for the use of diamond drills in this way. He showed that the costs per ton of recovered material were relatively equal for diamond drills and percussion drills, but that the diamond drill was more technically precise by a factor of 250% over percussion drills in six categories of cost effectiveness.

Various companies experimented with blasthole drilling using the diamond drill, but the two major mines where this type of work made a significant economic difference were Noranda in Quebec and at the Sullivan in Kimberley. Longhole blasthole drilling was the exclusive realm of the diamond drill well into the 1950's, but regular blasthole drilling was dependent on the economics of the mine operation.

C.D.M. Chisholm, the assistant mine manager at CM&S's Sullivan Mine, reported on the diamond blasthole work there in the October 1945 issue of the *Western Miner* magazine. He traced an increase in production from ten thousand tons in 1939 to four hundred twenty-three thousand tons in 1944 directly to the efficiency of this method. The contractor at the Sullivan was T. Connors Diamond Drilling Company with 11 drills on 2 shifts with more than 40 men employed. The average footage per machine shift was reported to be 86 feet. The cost for this method of drilling blastholes in 1944 was 40.1 cents per foot, with 23.6 cents in labor costs, 11.4 cents in equipment costs and 5.1 cents in diamond losses.

Not long after this time, CM&S took over all of its own underground drilling—moving the contracting companies out onto surface exploration. The permanent nature of much of the company's underground work and the increasing complexities of negotiating union jurisdiction in the mines helped force the company to assume responsibility for its own drilling. This had an important effect on the contractors who were now forced to be much more competitive to survive in a world consisting primarily of short-term exploration jobs.

An interesting sidelight to the blasthole drilling at the Sullivan was the machine being used there by Connors.

Called the TNT drill, it was a unique machine and it is thought to have been used solely on that property and nowhere else. It was developed in Vancouver and built in the machine shop at Hoffar's boat works at Coal Harbour. J.K. Smit in Vancouver continues to make spare parts for the machines which are still in use.

Powered by a peculiar compressed-air motor, the feed mechanism is a long pneumatic cylinder using a piston attached to cables to advance the drill head in place of rods. The compressed-air engine was the design of an Italian immigrant who came to Canada by way of Australia in the late 1920's. Felix Holznick first came to the attention of the technical world in 1929 when he arrived in Canada with designs for one of the first complex automatic transmission systems for automobiles. A small company was founded to produce these transmissions and a prototype was submitted to General Motors in Detroit for evaluation.

The design was not suitable for General Motors use and the company soon folded its doors.

The closing years of the 40's and the early 50's was a strange time for the Connors Drilling company. The firm landed some major contracts such as the exploration at Pine Point in the Northwest Territories where John Elneff supervised a drilling program continuously for the next twenty years. During the next year, 1949, Connors lost two of its founding partners. Norman Allan, one of the industry's most respected field supervisors, died of leukemia in January. Three months later, president Pete Murphy was dead. Looking very tired one April morning, he was persuaded to take the day off and rest. The next morning, the company received work that he had died at home.

Dunc Chisholm, the general manager of the firm, took over Pete's duties as president and led the company until 1968. He guided the company through some of the busiest times in the industry and through some of the leanest times Connors Drilling had to face.

The early 1950's were busy years for everyone in the industry. There was considerable activity again in the East and West Kootenays and Jim Cullinane, living in Nelson re-established an office and shop for Connors there. In addition to minerals exploration, the company was involved with the construction of the Waneta Dam on the Pend 'Oreille River ten miles south of the smelter at Trail. Besides the regular grout and test holes, Connors drilled over 125,000 feet of blast holes. Rudy Miller, who went on to become one of Connors top executives, came to work for Jim Cullinane on the Waneta Dam project.

In 1953, Connors negotiated a contract with American Metals, Climax Inc., for exploration drilling of their holdings at Little River in New Brunswick. This property later went into production as Heath-Steele Mines. The first drills for this work came from the Nelson shop, but it immediately became apparent that the ore body was of major significance. The company shifted most of its top field crews east and within three weeks, there were eighteen rigs in operation on the property.

When the Climax work came to an end, Connors Drilling entered into other contracts with various mining groups in the east. For a time, the office at Newcastle, New Brunswick, was the company's busiest branch. The new work took Connors crews to Red Lake and Sudbury in Ontario and into Quebec. Within a short time, another branch was developed at Senneterre, Quebec, and the company's head office shifted from Vancouver to Toronto. Diamond drilling remained at a low ebb in B.C. for some time to come.

The basic structure of the Connors company began to change in the mid-50's. From its beginnings in 1926, the T. Connors Diamond Drilling Company Ltd. was a loose

169

partnership of expert drillers and technicians who shared responsibility for the company's financial and operational decisions. With the deaths in previous years of a number of the original partners, their shares in the firm had passed to their heirs. The heirs were not themselves diamond drillers and new means had to be found to share responsibility for the operational decisions of the firm in order to assume the long-term debt necessary to purchase new equipment and expand.

Following considerable negotiation, the remaining partners arranged the sale of the firm to business interests in Toronto in order to reactivate the dormant shares. The Toronto group gave company management a free hand in operations and a good deal of practical advice which helped the firm survive the transition to a corporate structure, and to become more competitive in the contemporary marketplace.

IV

Some of the more unusual diamond drilling jobs to be done in modern times were undertaken by the two prominent Western Canadian firms: Boyles Bros. Drilling Company Ltd. and the T. Connors Diamond Drilling Company Ltd. The most spectacular of these was undoubtedly the removal of Ripple Rock.

The problem of Ripple Rock probably occupied more newspaper space in British Columbia during the first half of the 20th Century than any other natural phenomenon. The controversies that brewed about this great navigational hazard polarized political feelings on the coast and had far-reaching consequences for the growth of the province.

Beginning in the late 1800's, Ripple Rock became infamous as the graveyard of coastal shipping. Situated in Seymour Narrows roughly 120 air miles north of Vancouver, the Rock sat in the middle of the ship channel between Maud Island on the east and Vancouver Island on the west. Maud Island is a small island which lies in the western curve of Quadra Island not many miles from Campbell River, B.C. The channel at this point is less than one-half mile wide. Quadra is the last major island in the Inside Passage from Alaska for ships traveling south before they enter the spacious protected waters of Georgia Straits. Ripple Rock therefore was dead center on the primary shipping route serving the west coast of Canada.

The Rock itself consisted of twin peaks lying on a north-south axis which reached to within 10-feet of the water's surface at low tide. The Rock's dimensions were roughly 175-feet by 250-feet. At high tide, the 15-knot tide race smashed against the Rock and broke back upon itself creating the vicious whirlpools 40-feet in diameter which diverted even the most powerful steamships toward the Rock. Small craft such as pleasure boats, tugs and fishing vessels simply

171

disappeared into the frenzy to be shattered into pieces. All shipping had to stop and wait at either end of the channel for slack tide when it was safe to pass the Rock.

From 1875 until the early 1950's, it was reported that more than 24 large ships and over 100 smaller vessels were drawn into the Rock with a loss of lives totalling more than 114 people and millions of dollars of cargo. Many more millions of dollars were lost each year in time as 2,000 steamships and 7,000 small boats sat waiting in the Narrows for slack tide. None of this went unnoticed by the press.

In 1905-06 the first great debate began in the province's newspapers concerning Ripple Rock. A proposal was made at that time to use the Rock as a footing for a major railway bridge that would connect Vancouver Island with the British Columbia mainland by way of Quadra Island, Reed Island and Cortes Island. The plan was not a solution to the problem of the Rock as a menace to navigation, but was proposed as an alternate method for the distribution of goods and passengers to the mainland from the great ports that would arise from the shores of Vancouver Island. In this way, the large ships could avoid the Narrows altogether. The engineering problems and the astronomical financial requirements of this project soon put an end to the debate—but not before most of the coast's most respected personages made their case for or against it.

It was not until the 1930's that the problem of Ripple Rock developed into yet another brawl over the railway bridge. Some of the same people who argued with one another in 1905 were still around and could not resist the opportunity to jump back into the fray. Reason prevailed with just a bit of help from the Depression years and the conclusion was drawn that the whole idea of a railway bridge would have to be shelved until the province had hundreds of millions of dollars available to develop a new technology to cope with the water currents of the Narrows.

B.C. was stuck with the Inside Passage as its main transportation route north and stuck with Ripple Rock dead center in the Narrows. It was then that the first practical argument developed over removing the Rock altogether. For the next two decades the controversy raged across the pages of the Vancouver *SUN* and *PROVINCE* and the *Victoria TIMES*. There were blow-by-blow reports of ship sinkings and navigational problems around the Rock and more strident calls for its removal.

In 1943, an attempt was made to remove the Rock by attacking it from the surface. A large diamond drill mounted on a $160,000 barge was anchored over the Rock by steel cables attached to five cement blocks weighing 125 tons apiece. In the first 24 hours of drilling, one cable snapped and the entire barge was almost lost to the vicious currents swirling around the Rock. The project lasted an entire summer with cables breaking right and left and the currents lifting the barge out of position every time it drilled more than a few inches into the Rock. In the fall, the project was written off as a failure.

The U.S. government was anxious to see the Rock removed because it was delaying wartime convoys which used the Inside Passage, and so the problem of the Rock got widespread coverage in the States.

In 1945, the Rock was again attacked from the surface by another barge anchored overhead by another quite different cable system. The elaborate cable anchor worked somewhat better than its predecessor, but still there was so little progress made that this project, too, was written off to failure.

In 1949, the newspapers launched yet another campaign to call attention to B.C.'s problem Rock. In 1954, an American drilling firm, DeLong Corporation of New York, offered to do the job to demonstrate the abilities of a new company invention, a mobile drilling platform that was proving a great success in drilling for oil offshore in the Gulf of Mexico. They proposed to do the job on a "no-cure, no-pay" basis. The co-inventor of the drilling platform, Felix Schlickeisen, had intimate knowledge of the problem with Ripple Rock. Before moving south to work the Texas oilfields, he had been manager of the Silver-Skagit Logging Company and at another time the operator of a barge-towing service which used the Inside Passage. It was pointed out that a DeLong platform had just successfully withstood a 100-mile-per-hour hurricane in the Gulf. The offer was well-publicized, but the Canadian government backed off in hopes that one other, more logical solution might pay off instead.

As early as 1942, the Boyles Bros. firm had offered the government a plan to remove Ripple Rock for $800,000. The idea was to tunnel underneath the Rock and demolish it with explosives from below. Charles Hopper, a Boyles consultant and later to become the first president of the Canadian Diamond Drilling Association, was the man who first worked out the idea. There was some hesitation concerning the safety of this approach until it was explained that a core sample could be extracted by diamond drilling along the route of the proposed tunnel into the heart of the Rock. If the core proved that the rock formation was free from faults and non-porous, then the tunnelling method would appear to be the one feasible method of planting sufficient explosives to blow Ripple Rock out of the water.

In August of 1953, the National Research Council awarded a contract to Boyles to drill the exploratory hole from shore. The hole, collared on Maud Island, descended under the channel at an angle and then was deflected up into the heart of the Rock by wedging. Forty wedges were used to obtain the required path of the exploratory hole. In this exploration of the Rock, Boyles crews drilled more than six and one-half miles of core.

The cores proved that tunnelling would be feasible and work on the actual demolition project was supervised by the chief engineer of the Harbours and Rivers Engineering Branch of the federal Department of Public Works. The firm of V. Dolmage & E.E. Mason, Geologists and Mining Engineers, was in charge of the work for the government. Boyles Bros. Drilling was a major contractor on the demolition job along with Northern Construction and J.W. Stewart Ltd. Murray Smith, today the proprietor of Arctic Diamond Drilling of Whitehorse, was a Boyles driller on the project.

A shaft was sunk to a depth of 570-feet beneath Maud Island to allow at least 100-feet of solid rock overhead where the tunnel passed under the channel. The tunnel was driven west more than 2000-feet to a position under the north peak of the Rock. A 400-foot tunnel was driven from there under the south peak. Two vertical raises of 300-feet were driven up into the Rock peaks and a system of small box-hole entries and coyote drifts were made at selected locations.

Elaborate safety precautions were taken to protect those working underground and to protect the explosives from the dampness. The rock mass which had to be removed from the two peaks to permit placement of enough explosives exceeded one quarter of a million tons. The contractors used ammonium nitrate explosives set off with a high-velocity detonating fuse to avoid the problems of electric detonation of a mass this size in a wet atmosphere.

The blasting of the pinnacles was instantaneous and occurred on April 5, 1958. It was the largest non-atomic blast in history and the largest intentional underwater blast ever made. It remains something of a standard of comparison for describing non-atomic explosions. In 1966, when the U.S. Air Force dropped 1.4 million pounds of bombs on the Mu Gia pass in North Viet Nam, the explosion there was rated by newspapermen in comparison with the Ripple Rock blow: Ripple Rock was twice as powerful. Following the detonation, the rock mass in Seymour Narrows reached no more than 47-feet below the surface at low tide and since that time, the tides have carried much of the loose rock debris away.

Less than ten years later, Boyles was called upon to perform another job which involved unusual engineering precedents. The International Minerals and Chemicals Corporation decided to develop a fabulous potash deposit that had been discovered by oil drilling crews in the 1940's. At the time of discovery, the technical problems of mining the deposit were so overwhelming that nothing further was done there. But in 1964, the world's supply of potash which is a major ingredient in plant fertilizers was running low. Potash is potassium chloride which precipitates out of salt deposits left by the evaporation of the vast inland seas of prehistoric times. The deposit near Esterhazy, Saskatchewan, lies in a belt about 450 miles long by 50 miles wide at a depth of almost 3,000-feet below the earth's surface.

Sinking a mine shaft to the potash deposit was one of the most complex ventures ever undertaken by a resource firm in the western hemisphere. The deposit was covered by 289-feet of glacial till, 2,726-feet of water-bearing shale and limestone and 100-feet of rock salt. Inside the shale and limestone layer there was a 200-foot band of fluid quicksand under explosive pressures of 475 pounds per square inch.

It took more than one year to drive the shaft down to the infamous 'Blairmore' quicksand layer and another four months to go 200-feet through that extraordinary mess. The successful engineering solution saw the Boyles drillers drive holes around a 50-foot diameter area of the 'Blairmore' which were cased against the pressure of the quicksand. Refrigeration pipes were lowered into the holes inside the casing to freeze the quicksand at -60°F., turning the area into a 3-million-cubic-foot icecube. Using small jackhammers, construction crews chipped away at the ice/sand cube and worked their way down through the morass. A steel-plate cylindrical shaft was constructed as they went along to isolate the mine shaft from the quicksand layer. The miners and construction crews worked at temperatures of -34°F. in the hole, because of the refrigeration system.

The potash level itself was reached in June of 1962 and the mine has been one of North America's biggest producers of that vital commodity ever since. The IMC potash is a major component of COMINCO's 'Elephant'-brand fertilizers.

At roughly the same time that Boyles drillers were completing their job in Saskatchewan, Connors drillers were involved in an unusual exploration project for COMINCO in Greenland. For a number of years, COMINCO had been helping to direct the western hemisphere operations of the Danish mining firm Vestgron on the west coast of Greenland at Umanak. A potentially valuable deposit of lead-zinc was located at Marmorilik at latitude 71° North. The deposit was inside a rock bluff rising sheer from the side of a fjord which plateaued at about 3,000-feet elevation. A Connors crew led by Don Marshall was dispatched to Greenland to explore this deposit. The men lived aboard the former Canadian Arctic patrol vessel, *C.D. Howe*, which had been converted to civilian use as a floating exploration camp. Each day, it was necessary to climb to the top of the cliff carrying the day's drilling supplies. For the most part, these consisted of 50-lb. bags of salt which was used in solution as the drilling medium in permafrost. Once the basic deposit location was established, a camp was built on the plateau itself.

Surface drilling indicated an ore body with reserves of more than 2.5 million tons of lead-zinc, and plans were developed to mine the deposit. A subsidiary company was formed by COMINCO and Vestgron and named Greenex A/S. A mine shaft was punched into the face of the cliff at about the 2,500-foot level of elevation and a cable-car system built to

177

connect the mine portal with the base camp which was constructed at the foot of a scree slope across the fjord. Altogether a spectacular operation for the Arctic.

Another Connors team was sent to Marmorilik as soon as the mine shaft was begun, to follow the miners underground and continue to map the deposit. By 1971, there was over 2,000-feet of tunnel completed and 12 diamond drill holes finished. The core samples revealed twice the suspected reserves, over 4.5 million tons, with ore grades of better than 20% averaging 15% zinc, 5% lead and 1-ounce of silver per ton. The mine has been in operation since the fall of 1973 and has a life expectancy of more than fifteen years. The production averages 1,000 tons per year which is smelted in Europe.

With the work in Greenland in the late 1960's, Connors Drilling began to make an international name for itself just as Boyles had done in years past. The early 1970's saw Connors Drilling teams running exploration programs at Rubiales in the Pyrenees in Spain, in the Arab Republic of Yemen and in other foreign locations.

V

During the 1950's, the Boyles Bros. continued their development of foreign markets by incorporating new companies in South Africa—just down the road from the Boart diamond company in Johannesburg—and at Kitwe in Northern Rhodesia. In 1956, the company changed its status from a private to a public company. A branch was also established at Moncton, New Brunswick. By 1961, the company was reporting gross revenues from manufacturing and contracting and diamond product sales in excess of $7 million.

Certain changes in technology began to have an effect on the industry. Long-hole percussion drills began to appear on the market in the late 1950's, and the economics of blasthole drilling swung once again to the improved percussion drills because of escalating wages and the high cost of diamonds. Perhaps the major challenge to Boyles supremacy in the manufacture of diamond drilling equipment came when the American-based E.J. Longyear company introduced the wire-line core barrel system at the end of the decade. Longyear had always been the major manufacturer of diamond drill equipment after the Sullivan Machinery Company had failed to keep up with 20th Century drill technology. Chicago Pneumatic Tool Company was also a strong contender in the marketplace in the late 1950's with its CP-55 and CP-65 underground drills, which saw widespread use in the mines of Africa.

The wire-line system was another landmark in diamond drill technology. A coupling device mounted at the end of a light cable is dropped down inside the string of drill rods to latch onto the inner tube and releases it from the outer barrel. The inner tube containing the core is then hoisted to the surface. In this way, the driller can recover the inner barrel containing the core sample without lifting the string of drill rods from the hole. This drastically reduced the costs of drilling and the % core recovery.

At this same point in time, many diamond drilling contractors were being forced to curtail their operations due to the rising cost of diamonds and other supplies, and the increasingly fierce competition in the industry. Inspiration Mining and Development Company of Montreal, founded during the Depression years, was the major contractor at such projects as the Beaucage Mines in Quebec. In 1956, the company was producing over half-million feet of core from 66 drills employing 464 people. The next year, business had fallen off drastically and the company did just 300,000 feet of core with 27 drills and 228 employees. In 1958, contract footage was down another 22% and the average price per foot for their work fell from the 1957 high of $3.67/foot to $2.25/foot. In 1961, the company shifted the emphasis of its operation to general contracting and manufacturing.

The 1960's brought even more drastic changes to the industry with the news that the Boyles Bros. Drilling Company was sold to Inspiration, which had become the fifth largest general construction contractor in Canada. The merger took place in 1966, most of the Boyles operations were moved east to Ontario, and the company name changed to Boyles Industries.

In January of 1968, the Inspiration organization announced that it was closing all of the Boyles' Vancouver operations. Many of the Boyles employees struck out on their own rather than make the move east. Eight employees from the manufacturing plant, led by W.A. Rennison and Carl Wahlstrom, formed the Wesdrill Equipment concern in June of 1968 to maintain a diamond drill manufacturing concern in the west. The company began producing diamond bits, but since that time they have introduced a new drill to the industry which represents the next step forward in diamond drilling technology.

Sudden drastic changes in the fortunes of the Inspiration company sent the transplanted Boyles subsidiary into receivership and within a year, they were no more. The manufacturing arm of the Boyles concern was sold to Dresser of Texas and the contracting business was eventually repatriated to the west by Connors Drilling, but not before many of the old Boyles hands either retired or opened their own small contracting firms. The Boyles Bros. drills are still produced in England and at Orillia, Ontario in Canada by the Dresser Company under the name of Boyles Industries.

At the moment in time when the Boyles Bros. Drilling Company came to its end, the firm of Connors Drilling Ltd. was undergoing something of a renaissance. The Connors company moved its head office back to Vancouver from Toronto in 1960, anticipating an accelerating market for exploration drilling in the West. At the time, the company was fielding more than one hundred drillers during the peak summer season.

Some of the company's top executives got their start in diamond drilling earning 'summer dollars' as drill helpers, just the way Jim Cullinane, Sr., did in 1925. One 'summer dollar' employee was R.R. "Bob" Carver, a Vancouver native who had worked for Connors while working toward his degree as a mining engineer at the University of British Columbia. After university, Bob moved to the United States and made a fine reputation for himself through twenty-years with the giant contracting firm of Sprague & Henwood of Scranton, Pennsylvania. Bob kept in close contact with the Canadian drilling industry, often using Canadian drillers on projects he organized for Sprague & Henwood overseas.

In 1966, he returned to his home town to become operations manager for Connors Drilling. In that post, he created the streamlined corporate structure that the company uses today. He guided Connors through negotiations leading to the sale of the firm in 1968 to Bow Valley Industries of Calgary, Alberta. Bow Valley was interested in oil-well drilling and resource development. H.G. "Red" Bryden came to Connors to help integrate the company with other Bow Valley affiliates including Hi-Tower Drilling Co., Bow Helicopters and later, Wesdrill Equipment Ltd.

H. "Bert" Cameron came to the company as Bob Carver's executive assistant and company management launched on a modernization and expansion program that has made Connors Drilling the largest diamond drilling firm in Canada. The program of acquisitions was engineered by Donald R. Seaman, senior vice-president of Bow Valley and president of Connors, Red Bryden, Bob Carver and Bert Cameron.

Connors acquired a U.S. branch with the purchase of Malcolm McPherson Drilling Inc. of Montrose, Colorado. Men with many years of drilling expertise in other parts of Canada came to Connors with the acquisitions of Griffith Brothers Drilling of Lac du Bonnet, Manitoba. Throughout the late 1960's and early 1970's, the former Connors executive officers, Jim Cullinane and Dunc Chisholm, worked with the new management team to ensure that the company built upon solid foundations for the future.

With the sudden passing of Bob Carver in 1972, Bert Cameron became vice-president and general manager of the company with Red Bryden his administrative assistant. The program of acquisitions continued with the 1974 purchase of Inspiration's contracting services in western and central Canada and of Inspiration's Thetford Mines branch in Quebec. The company spent the next year reviewing its technological expertise, modernizing its equipment and restructuring its twelve branches in Canada and the U.S.

With 276 drill rigs drilling nearly 1.5 million feet in 1975, Connors has established itself as Canada's largest diamond drill contractor of modern times. The Longyear company's contracting arm has also made its presence felt in the Canadian marketplace, and many smaller companies are thriving here as well. Some of the small contractors find that staying small is actually more profitable than competing with giants like Connors and Longyear. Many of the small contractors in Canada at the present time are owned and operated by drillers who have worked for one or more of the major contractors at some time and acquired a reputation within the industry for having expertise in one special aspect of drilling or another. The presence of these firms keeps the competition strong for the major contractors.

Following the upheavals of the past decade, the diamond drill industry of the 1970's is in relatively good health. There are a number of indicators which serve as guides to the fortunes of the industry. Sources within the world of diamond supply report that industrial applications now require 55 million carats of the stones each year, and that more than 10% by weight of this demand comes from the drilling supply industry. Expectations of this demand growing to exceed 100 million carats by the end of the decade, with an even greater % share going to the drilling industry, give cause for optimism within the industry. In an effort to insure that diamond supplies are adequately maintained for use by the mining and exploration industries, COMINCO and other industry giants have acquired partnerships in the diamond mining and distribution businesses.

The diamond drilling industry is also beginning to prepare its own future by sponsoring, through the Canadian Diamond Drilling Association, a number of training courses across Canada designed to provide the industry with drill helpers who are technically proficient at their work. The CDDA Winnipeg chapter conducted the pilot program in the mid-60's with the support of the Griffiths Bros. firm and Midwest Diamond Drilling Ltd. Today, various chapters have arranged training courses in conjunction with the provincial manpower services. A diploma program for Assistant Diamond Drilling Operators is offered with CDDA sponsorship at the Northern College of Applied Arts and

Technology at Haileybury, Ontario, and an up-grading course for drill helpers is given by the CDDA in Vancouver.

In contemporary times, the diamond drill has become a much more sophisticated machine than could be anticipated even ten years ago. This is partly the result of the integration of more sophisticated technologies in use by the resources industries, and also due to a demand that the diamond drill play a more diverse role in life outside the mining industry.

The diamond drill today is used to perform a multitude of functions besides core sampling for geological applications. One important use is in delivering a neat, straight hole to a particular point for a specific purpose: for example, the diamond drill is used to deliver grout and cement mixtures to rock fractures deep underground to prepare solid and water-tight foundations for the construction of bridges and dams and other structures. The diamond drill has also been used as the handmaiden of mercy to provide rescue workers with a means of delivering air and food and water to the victims of mine disaster trapped underground.

Of course, there have also been occasions when diamond drills have been used for nefarious purposes. Stories are told of firms who rented idle equipment to 'construction' people only to be informed by very irate police officers that their customer had 'constructed' neat holes through a bank vault.

One enterprising criminal who rented a drill for such use went and stole the bits to go with it. He had some acquaintance with the industry and understood that he could return the bits for the credit of the diamonds. Unfortunately, he did not know that bit manufacturers issue positive proof of ownership to prevent such a thing and he could not provide this proof when he tried to return the bits for credit. His mistake cost him 10-20 years and the profits of an otherwise successful bank robbery.

The scientific establishment has long enjoyed the use of the diamond drill. The geology department at the University of British Columbia owns its own drill for use in field experiments on rock structures, to provide a means of delivering sensitive instruments into the earth to measure stress for example. The diamond drill has long been a staple tool of scientists working on projects to plot the earth's 'mantle'— that portion of the earth's interior which separates the crust on which we live from the molten core of the planet.

As part of the International Upper Mantle Project in 1964, Canadian drillers working under contract for the Department of Mines and Technical Surveys drilled a hole more than 4,000-feet deep in the high Arctic to locate the beginning of the mantle. A total of more than 10,000-feet of hole was drilled at Muskox Intrusion near Coppermine in the Northwest Territories. The drillers worked in temperatures of -30°F. with 30-mph winds. The excellent core samples, with a recovery rate exceeding 98%, were later turned over to the Canadian Geological Survey for their use.

Since 1973, scientists from nearly a dozen nations have been engaged in a massive research project in Antarctica. Martin McGale of Canadian Longyear was superintendent of diamond drilling at the Dry Valley Drilling Project near Ross Island (N.Z.). Additional drills were at work near McMurdo Base in the U.S. zone.

It was announced this year that these scientists are planning to drill through the 1,400-foot thick ice cover of the Ross shelf to reach a body of water which has been cut off from contact with the outer world for more than one million years. They plan to lower nets and traps and cameras through diamond drill holes to study "the only remaining community of organisms left undisturbed since the pleistocene"—the Great Ice Age. They even hope to take core samples from the seabed more than 780-feet below the bottom of the ice layer to study the ancient history of the Antarctic continent.

Diamond drills are sometimes confused with their massive cousins the oil-well drills which sometimes use diamond-studded bits as well. The two different types of machines are used to locate very different types of energy resources such as gas and oil with the big drills, coal and uranium with the diamond drills. Currently there is a project underway at Coso Springs in the California desert on the grounds of the China Lake Naval Weapons Center, in which a diamond drill is being used to provide scientists and engineers with data about yet another natural energy source—geothermal power. The program, funded by the U.S. Energy Research and Development Administration, is aimed at pinpointing areas in the earth's crust where volcanic activity produces 'heat wells' which might be exploited commercially as a source of power for the future.

The drill being used on the project is equipped with blowout preventers—just like an oil-well rig—to withstand pressures of 5,000-lbs. and temperatures of 600°C. It is the new Wesdrill Model 60 manufactured by Wesdrill Equipment Ltd. of Richmond, B.C. The Wesdrill represents a new generation of diamond drills which advance the technology of this industry yet another step forward. The drill incorporates all the latest advances in mechanical operation including a hydraulic swivel head and clutch and disc brakes on the hoists, a true eight-speed transmission, high-speed hydraulic chuck and 1600 RPM drive tube—but its major departure from standard rigs is its instrumentation.

The basic instrumentation includes analysis of the rate of feed, the true weight that the bit is carrying, speed of bit and drill rod torque. Many additional devices can be installed to automate the equipment almost to the point of computerization. The basic instrumentation can be preset by the driller to react to changes in rock structure and record this data on graph machines. One of the first applications of the Wesdrill was by Connors Drilling on a job in the Colorado oil shales. Six different rigs had met with failure to provide a useable core sample from the broken ground. Connors was the first to provide this core, because of the sophistication of the Wesdrill's control system. Even in areas where complete core recovery was impossible, the graphic analysis of changes in the rock structure provided by the machine's instrumentation gave the geologists the information they required.

The Wesdrill people readily admit that they could carry the automation of their machines to much greater extent, but they understand why the industry is not yet ready for this. A new generation of drillers accustomed to working with advanced electronics instruments is only now beginning to appear on the scene. Further advances wait for more technologically proficient personnel.

There is as well a problem of money. Many of the smaller contractors just cannot afford the sophistication of machines like the Wesdrill. The margin of profit in the industry remains so competitive that most firms simply cannot afford the kind of automation that has become commonplace in industries such as petroleum exploration. Some clients of the industry, aware of the advantages of having the data provided by the new instrumentation, are asking their contractors to provide these additional services and are making allowances in their contracts for the change-over to this new technology. In this manner, the diamond drilling industry is slowly taking the next step to a higher degree of craftsmanship.

The future, as always, remains a surprise, but there are some indicators which give a hint of what tomorrow may bring. Various experiments are being conducted with different types of technology which could be applied to drilling. The use of sonic waves to shatter rocks could displace the use of drilling blastholes and demolishing rock with explosive charges.

During the building of the St. Lawrence Seaway in the late 1950's, a jet-flame 'drill' was used to remove large quantities of certain types of rock. The U.S. government is experimenting with thermal boring devices. Descendents of a concept developed in *Tom Swift Jr. and his Atomic Earth Blaster*, the thermal devices (called *subterrenes*) owe their practical invention to half-facetious suggestions made at a scientific congress in the early 1950's by the Russian scientist Dr. Vladimir V. Beloussov. These devices use solid heads of laminated metals to convert electricity to heat energy. A prototype used in experiments at the Los Alamos testing grounds bored a 2-inch diameter hole through granite at the rate of 30-inches per hour. Operating at temperatures exceeding 2200°F. and with a power source equal to just thirty 100-watt lightbulbs, the subterrene left a shaft with smooth walls of an obsidian-like glass which supplied the structural component usually provided by casings. Using atomic power, larger machines could bore shafts to any size or shape—round, square, triangular and so on.

The advent of such machines will reduce the use of diamond drills to ever more specific purposes. While some people still contend that there will always be a need for the real, solid physical core of rock to touch and examine and test, there are those who quietly contend that someday the diamond drill will go the way of the slide rule and for much the same reasons. Some of the industry's old-timers who have watched the world change in so many exciting ways say they would not be surprised to learn of the eventual development of some sort of true 'x-ray' device which will give quantitative and qualitative information about rock structures at virtually any depth, without physically disturbing the surface of the earth at all. They point to the instrumentation of machines like the Wesdrill as evidence of that trend.

For now, diamond drilling remains the most efficient means of solving so many puzzles that we can look forward to depending upon this industry for a long time to come. Whatever the future brings, it develops from here—from this moment in time and this level of technology. Just as the

Wesdrill is a descendent of Leschot's concept, so too it may be the ancestor of devices that we can not yet imagine.

With luck, the diamond driller will survive to follow this technology into the future. The job description may change, but we should hope that the people will not. There will always be a demand for people with the curiosity, the independence of spirit and the high sense of craftsmanship which the diamond drillers exhibit to help provide an increasingly more complex world with the resources it needs to survive and flourish.

Sources

It has been the author's pleasure to meet and work with many people in the course of researching and writing this history of the diamond drilling industry in western Canada. My most important sources of information for this book have been the people who work in the industry. I would like to give these people the credit due them in assisting me with my work:

JAMES A. CULLINANE—a fine Irish gentleman, a storyteller and historian who shared with me many days of his time and experience and provided an excellent selection of photographs for this book. It was a rare pleasure for me to work with Jim Cullinane and without his recollections this book would not have been written.

D.C. "DUNC" CHISHOLM—another fine man who spent his life as part of the diamond drilling industry. As the former president of Connors Drilling Ltd., Dunc contributed a great deal to the history which is recorded here.

JOHN W. BOOTH—for almost forty years, John Booth was a prime mover of the technical staff of the Boyles Bros. Drilling Company Ltd. I am particularly grateful for the material he provided concerning the old Boyles company which has now passed into history and I have relied upon his own writings in my interpretation of the earliest days of diamond drilling.

DONALD R. SEAMAN, president of Connors Drilling Ltd.

H. "BERT" CAMERON, Connors' vice-president and general manager

H.G. "RED" BRYDEN, Connors' administrative manager and current president of the Canadian Diamond Drilling Association.

These gentlemen initiated this book project as part of the celebration of Connors' 50th anniversary as a diamond drill contractor. They introduced the author to all of those who have contributed so generously to this book and they gave me a great deal of their time and professional expertise to ensure the success of this history.

My thanks to all the employees of Connors Drilling Ltd. who provided me with anecdotes and stories, technical information and photographs of the activities of both the Connors firm and the Boyles Bros. Drilling Company. I would most especially like to extend my best wishes to:

R.V. "RUDY" MILLER

GERRY BEDARD

R.L. "BOB" ROBERTSON

ARNE ROSEN

JOE BENARD

JAMES J. CULLINANE

...and to CLAIRE McDONALD and the office staff of Connors Drilling who helped proofread portions of the manuscript and gave me encouragement in my work.

W.A. RENNISON and WESLEY SMYTH of Wesdrill Equipment Ltd. provided me with a great deal of valuable information about technical aspects of diamond drill equipment and read the manuscript.

BRUCE ROZENHART of COMINCO provided me with the photographs of the Black Angel Mine in Greenland and helped me obtain other information used in the book.

GEORGE BRANDAK and the staff of the Special Collections division of the University of British Columbia library system assisted me in researching the general historical portions of the book and provided me with a great many of the photographs used here.

The staff of the Vancouver Public Library gave generously of their time and effort to assist me in my research. I especially wish to thank the staffs of the Business division, Northwest History section and the Historical Photographs division for their help.

The staff of the Vancouver City Archives assisted me with my initial research and generously allowed me to publish quotations from archival materials.

As explained in the FOREWORD to this book, it was not envisioned as an encyclopaedic source of information about the diamond drilling industry. Rather, it is the first slim volume to scratch the surface of an enormously rich and varied vein of technical and social history that should be mined before time places the original sources out of our reach. There is at present no formal collection of informaton about the industry, but those in possession of such documents and photographs and stories should contact the Canadian Diamond Drilling Association with their materials. An historical collection of mining hardware is being assembled at the mining museum at Britannia, British Columbia.

Published Sources
BOOKS

DIAMOND DRILL HANDBOOK (3rd edition), James D. Cumming and A. Percy Wicklund, 1975, J.K. Smit & Sons Diamond Products Ltd., Toronto, Ontario.

THE DRILLING OF ROCK, K. McGregor, 1967, CR Books Ltd., London, England.

ROCK DRILLING, Richard T. Dana and W. L. Saunders, 1912, John Wiley & Sons, New York.

DIAMOND DRILLING FOR GOLD & OTHER MINERALS, George Alfred Denny, 1900, Crosby Lockwood and Son, London, England.

THE ROMANCE OF MINING, T.A. Rickard, 1944, Macmillan Company of Canada 'Ltd., Toronto, Ontario.

THE HILLS OF ADONIS (A Quest In Lebanon), Colin Thubron, 1968, William Heinemann Ltd., London, England.

OTHER

THE DIAMOND CORE DRILL—100 years old (1863-1963), John W. Booth, Boyles' *THE CORE BOX*, Feb. 1963. [courtesy J.W. Booth]

COL. F.E.B. BEAUMONT AND THE DIAMOND DRILL, John W. Booth, Boyles' *THE CORE BOX*, Nov. 1964. [courtesy J.W. Booth]

MACHINES THAT CHANGE THE WAY THE WORLD WORKS, James A. Drain, address to the 1966 "Pittsburgh Dinner" of the Newcomen Society in North America, Nov. 16, 1966. [courtesy Vancouver Public Library]

HELLGATE AND SANDY HOOK, *The Manufacturer and Builder*, New York, July 1872, Vol. IV #7 [courtesy J.W. Booth]

HISTORICAL OUTLINE OF B.C. MINING, William H. White, *Western Miner and Oil Review*, June 1961.

B.C. MINING—100 Years, John Warren, *The Financial Examiner*, 2 parts: Oct. 28, 1967 and Nov. 18, 1967. [courtesy Vancouver City Archives]

THE SAGA OF MINING IN BRITISH COLUMBIA, Bruce Ramsay, 1957, C.M. Oliver & Co., Vancouver, B.C. [courtesy Vancouver Public Library]

ROSSLAND—THE GOLDEN CITY, ed., 1949, Miner Printing Co., Rossland, B.C.

THE MINERS IN KOOTENAY, *Victoria Colonist*, March 10, 1900. [courtesy Vancouver City Archives)

JOE MORIS, Pauline Battien, Miner Printing Co., Rossland, B.C.

SLOCAN MINING CAMP, B.C., C.E. Cairns, 1934, Geological Survey of Canada, Dept. of Mines, Pub. #2358.

BLUEBELL—THE STORYBOOK MINE, Craig Weir, *Western Miner and Oil Review*, May 1960.

BOYLES BROS.: Elmore and Page, *and* FROM FARM TO FORTUNE, ed., Boyles' *THE CORE BOX*, B.C. Centennial Issue (1858-1958).
[courtesy R.L. Robertson]

CANADIAN TUNNELLERS AT GIBRALTAR, Major George F.G. Stanley, *Canadian Geographic Journal*, June 1944.

"Business Out West", editorial column, *Canadian Business*, Jan. 1945.

BLAST-HOLE DIAMOND DRILLING AT THE SULLIVAN MINE, C.G.M. Chisholm, *Western Miner*, Oct. 1945.

MERITS OF BLASTHOLE DIAMOND DRILLING COMPARED WITH PERCUSSION DRILLS, C.H. Hopper, *The Northern Miner*, Dec. 11, 1948.

RIPPLE ROCK, ed., Boyles' *THE CORE BOX*, Oct. 1953.
[courtesy R.L. Robertson)

THE MYSTERY OF POTASH, *Cominco Magazine*, Sept. 1964. (condensed from CPR *Spanner* magazine)

DIAMOND DRILL HITS "BOTTOM" IN THE UPPER MANTLE PROJECT, *Financial Post*, March 7, 1964.

LAKESIDE TRAINING SCHOOL MAY BEAT DIAMOND DRILLER SHORTAGE, *Financial Post*, Nov. 27, 1965.

THERMAL BORING DEVICE MELTS AWAY GRANITE, Walter Sullivan, *N.Y. Times*, Dec. 25, 1971.

RENEWED INTEREST IN KING SOLOMON'S GOLD MINE, ed., *The Northern Miner*, August 5, 1976.

...and numerous short items from a variety of magazines, newspapers and reference books. The trail of general information about the diamond drilling industry is sometimes hard to follow, but I hope the sources listed here may give other writers and interested readers a place to start. Best wishes!

Dan Fivehouse
Vancouver, 1976